Concrete Pavement Design Guidance Notes

T0221475

Also available from Taylor & Francis

Reynolds's Reinforced Concrete Designer Handbook 11th edition
T. Threlfall *et al.*   Hb: ISBN 0–419–25820–5
                                   Pb: ISBN 0–419–25830–2

Reinforced Concrete 3rd ed
P. Bhatt *et al.*   Hb: ISBN 0–415–30795–3
                                   Pb: ISBN 0–415–30796–1

Concrete Bridges
P. Mondorf   Hb: ISBN 0–415–39362–0

Reinforced & Prestressed Concrete 4th ed
S. Teng *et al.*   Hb: ISBN 0–415–31627–8
                                   Pb: ISBN 0–415–31626–X

Concrete Mix Design, Quality Control and Specification 3th ed
K. Day   Hb: ISBN 0–415–39313–2

Binders for Durable and Sustainable Concrete
P.-C. Aïtcin   Hb: ISBN 0–415–38588–1

Aggregates in Concrete
M. Alexander *et al.*   Hb: ISBN 0–415–25839–1

Information and ordering details

For price availability and ordering visit our website **www.tandf.co.uk**

Alternatively our books are available from all good bookshops.

Concrete Pavement Design Guidance Notes

Geoffrey Griffiths and
Nick Thom

CRC Press
Taylor & Francis Group
Boca Raton London New York

CRC Press is an imprint of the
Taylor & Francis Group, an **informa** business

A TAYLOR & FRANCIS BOOK

Contents

Figures

Tables

Equations

Acknowledgements

We owe an immense debt to the staff at the University of Nottingham for their initial guidance in introducing us to the design of concrete pavements through their MSc course in Highway Engineering. We would like to thank the following staff for their interesting lectures around the subject of pavement engineering:

- Professor E. Barenberg for his visiting week of lectures in 1994;
- Andrew Dawson for his opinions on the design of pavements;
- Dr Kim Elliot for the quality of his notes on concrete design.

The book also owes a substantial debt to Britpave, the UK concrete pavement association, David Jones, the Director and the many industrial members of the group. In particular we would like to thank Graham Woodman of WSP, Bob Lane from BAA and Jonathan Green. Britpave has provided a stimulating forum for the debate of issues associated with concrete pavements.

I must also thank my fellow colleagues at Ove Arup & Partners who have, over the years, offered me the opportunity to work on many major prestigious civil engineering projects around the world.

We would like to thank the following organisations for permission to reproduce certain material in this text:

- TRL Figure 6.3, Table 6.6, Equations 6.5, 7.3, 7.4
- American Concrete Pavement Association Figure 9.7
- HMSO Figures 10.5, 10.6, 10.7

Notations

Chapter 4

$f_{t,sp}$ = tensile splitting stress
F = applied force
l = cylinder or cube length
d = cylinder diameter or cube length
$f_{t,axl}$ = tensile axial stress
$f_{c,cyl}$ = compressive cylinder strength
$f_{c,cube}$ = compressive cube strength
$f_{t,fl}$ = tensile, flexural strength
F = applied load
L = distance between supporting rollers
d = beam depth
b = beam width
$f_{t,axl}$ = mean tensile axial strength
hb = slab thickness
β = factor to adjust 28-day strength
S = cement factor, 0.25 for normal and rapid hardening cement, 0.38 for slow hardening cement
T = number of days since construction
E_{ci} = Initial Tangent Young's Modulus for dynamically loaded uncracked material
C = correction factor, specific to a material
E_c = static uncracked Young's Modulus (Secant Modulus)
$E_{long-term}$ = long-term static uncracked Young's Modulus
$f_{t,fl}$ = flexural tensile strength from the standard beam test
$\sigma_{t,fl}$ = tensile stress calculated under a design load
R = ratio maximum to minimum stress within the loading cycle
t = duration of loading pulse
N = the number of load repetitions to failure

Chapter 5

ℓ = radius of relative stiffness (m)

k = modulus of subgrade reaction, = stress/deflection (Pa/m; normally quoted as MPa/m)

E = Young's Modulus of pavement slab (Pa; normally quoted as GPa)

h = Thickness of the pavement slab (m)

v = Poisson's Ratio of the pavement slab

E_s = Young's Modulus of the subgrade, considered as infinitely thick (Pa; normally quoted as MPa)

v_s = Poisson's Ratio of the subgrade

P = applied load (N)

a = radius of contact area for the point load (m)

a_1 = distance from the load centre to the corner = $2^{0.5}a$

M_o = limiting moment of resistance of the slab per unit length

M_n = negative moment of resistance

M_p = positive moment of resistance

M_c = limiting moment of resistance for mass concrete slab per unit length

M_s = limiting moment of resistance for reinforced concrete slab per unit length

f_t = concrete tensile stress at failure

h = thickness of concrete slab

M_s = net resultant moment

a_s = net resultant area of reinforcement

M_t = transverse moment

a_t = area of transverse reinforcement

M_l = longitudinal moment

a_l = area of longitudinal reinforcement

M_S = limiting moment of resistance for centrally reinforced concrete slab per unit length

f_s = steel tensile stress at failure

a_s = net resultant area of reinforcement

h = thickness of concrete

M_{cc} = limiting moment concrete compression per unit length

d = effective depth of section

X = depth as described in Figure 5.5 and Equation 5.18

m = separation of dual wheels

t = separation of tandem wheels

E = Young's Modulus of concrete (MPa)

α = coefficient of thermal expansion (per °C)

T = temperature difference, top to bottom of slab (°C)

v = Poisson's Ratio

L_x = distance between joints in x direction

L_y = distance between joints in y direction

ℓ = radius of relative stiffness
h_1, h_2 = thicknesses of slabs 1 and 2
E_1, E_2 = Young's Moduli of slabs 1 and 2
M_1, M_2 = bending moments experienced by slabs 1 and 2, per m length
σ_{1t}, σ_{2t} = maximum tensile stresses in slabs 1 and 2 under applied moment M_0
M_0 = applied moment per m length

Chapter 6

F = 'damage factor' applicable to axle load P (also known as 'equivalence factor' or 'wear factor')
P = axle load in kN
P_s = standard axle load in kN
n = relative damage exponent, commonly taken as 4
h_{design} = design pavement thickness
h_{mean} = mean pavement thickness based around a calculation
Z = normal distribution adjustment
S = standard deviation based on the calculation and construction variability

Chapter 7

W_{18} = traffic loading in standard 18 kip (80 kN) axles
D = pavement thickness in inches
E_c = Young's Modulus of concrete in psi
k = modulus of subgrade reaction in pci
S'_c = the mean 28-day modulus of rupture – that is flexural strength
S_o = the standard deviation of the data used to construct the pavement
Z_R = values (for a statistically normal distribution)
J = values: joint factor
C_D = drainage coefficients
ΔPSI = change in pavement serviceability index during pavement life
P_o = initial serviceability condition
P_t = terminal serviceability condition
P_s = percentage steel reinforcement
L = slab length, the distance between free Edges (feet or m)
F = friction factor at the base of the slab, 1.8 for lime, cement or bitumen stabilised material, 1.5 for unbound gravel or crushed stone and 0.9 for natural subgrade
f_s = ultimate fracture stress of reinforcement used in the slab (psi or MPa)
f_t = indirect tensile strength, mean 28-day value used in design
S'_c = flexural strength (i.e. modulus of rupture), mean 28-day value
R_T = aggregate factor taken as: gravel = 5/8, crushed rock 2/3

Chapter 1

Introduction

1.1 Background

Concrete pavements are a frequently misunderstood form of construction. Many practising engineers are unaware of the basic design issues which should really be understood before undertaking the construction of any pavement in concrete. Most engineers receive extensive training in the design of reinforced concrete structures but very little formal training is given in the design of concrete pavements. Yet concrete pavements are designed quite differently from reinforced concrete structures.

This document presents an explanation of the various design methods, materials testing methods, specifications and details that are associated with construction of cement bound (i.e. concrete) pavements. The design of concrete pavements has to be treated in a different manner from that of bituminous or granular pavement systems. This book is written partly out of despair; UK pavement design lacks a clear accepted text defining the terminology and methods employed within the concrete pavement industry and much confusion currently exists in describing pavement materials, strengths and loadings. Of course, a single book can never be perfect, but this one is intended to be a working script that will assist an engineer and, where appropriate, point towards other references. The chief aim of this book is to summarise the main design methods in use in the UK and USA, although reference will also be made to methods elsewhere where they are seen as particularly appropriate. The book is clearly unable to draw together every piece of current information or opinion in the field; however, every effort has been made to be as accurate as possible within the confines of the need to produce a concise but informative script on the subject.

The book is not a code of practice or design method; engineers must use the original design standards precisely to produce pavement designs.

1.2   Standard design methods

A successful design method must include each of the following key elements:

- a standard description of loading;
- a technique for describing the subgrade support and the required foundation design;
- a method to determine pavement strength (commonly undertaken as a Westergaard-derived stress calculation);
- calibration; anchoring the method against data from pavement trials;
- a specification, to control construction materials and techniques.

A number of standard methods may be used to produce successful pavement designs. But it is important that practising engineers understand that the different design techniques cannot be mixed and matched. None of the design methods is a precise model of the real world; each method calculates pavement thickness in a different way. If sections taken from different methods are combined inappropriately, the basis of the design is lost. Thus, an AASHTO design must be produced exactly in accordance with the AASHTO method and constructed in accordance with the AASHTO specification. Mixing and matching design methods reduces the accuracy of a design.

1.3   Calculating stress

Another important reason for writing this book is that many young designers appear to think that substantial benefits may be obtained if only one can calculate the stress within the pavement. Analysing concrete pavements appears to be rather like the mediaeval alchemists' quest to turn base metal into gold; the quest can never be achieved. Experienced engineers realise that their calculations are imperfect and cannot ever exactly match real site conditions. Engineers should understand that their calculations are no more than estimates and artistic descriptions of pavement failure. All pavement engineering is a subtle blend of art and science, and concrete pavement engineering is no exception.

1.4   Economic viability of concrete pavement systems

Yet another excellent reason for writing this book is that concrete is commonly the cheapest form of construction for large heavily trafficked pavements. The economics of concrete pavement construction becomes even more attractive if large quantities of sand and gravel are present on site. Large areas of central Europe have vast reserves of sand and gravel but no hard stone aggregates. Concrete pavements are particularly attractive solutions to the construction of large infrastructure projects in these areas. It should also be noted that concrete pavements are often particularly appropriate in hot climates or situations where very low subgrade strengths may occur.

However, concrete pavements are currently suffering from a poor public relations image. Many members of the public have driven over poorly constructed jointed concrete motorways and perceive them as both noisy and bumpy when compared with bituminous surfaces. The bitumen industry has also supported a number of design guides that have improved the competitive perception of bituminous roads. This design guide is intended to assist practising engineers in ensuring that when concrete is considered it is used correctly and, if economically attractive, also constructed successfully.

1.5   A note on concrete

This is most certainly not a book about concrete. It is a book about the use of concrete in a particular application, namely pavements, and that means that it is only necessary to understand concrete as a material having certain relevant properties. Thus, the type and size of aggregate used, the cement and water contents, are only important here in terms of the way they affect the engineering properties of the finished product. There are numerous excellent texts, notably Neville [1], which deal in detail with concrete as a material. In brief, the key properties of interest here are:

- Strength – the stress required to cause fracture;
- Fatigue characteristics – the tendency to fail at a stress lower than the fracture strength if enough load applications (e.g. truck wheels) are applied;
- Stiffness – the ratio of applied stress to resulting strain;
- Expansion coefficient – the relative increase in dimension as temperature increases.

Of course, it is necessary to have a certain appreciation of how the material is produced (usually in a batching plant) and how it is incorporated into a pavement (either wet formed or laid dry and roller compacted). Similarly, the fact that all concretes take a certain time to gain strength has consequences which impact on design and specification, as does the fact that they inevitably shrink measurably during the strength gain process. Concrete materials will also shrink and warp as they cure or age. However, here it will simply be assumed that the appropriate techniques are employed during this 'curing' process to make sure that shrinkage cracks are avoided and that loading is not applied prematurely. The actual processes and the different curing rates induced by different types of cement and at different temperatures are not issues dealt with here.

1.6 Units

This book is written principally using the SI system of units. The majority of the calculations presented are performed using metres, newtons, pascals, etc.; imperial units (inches, pounds, psi, etc.) are only used when alternative metric units would clearly cause confusion, which in practice means when illustrating certain American computational approaches.

1.7 Reference

1. Neville, A.M., *Properties of Concrete*, 4th edn, Longman Scientific and Technical, New York, 1995.

Chapter 2

Surface slab systems

2.1 Description of systems

Surface slab systems, otherwise known as 'rigid' pavements, are the most easily recognised form of cement bound pavement construction. Three alternative and interchangeable forms are commonly found in civil engineering projects:

- *mass concrete slabs or unreinforced concrete (URC)*;
- *jointed reinforced concrete (JRC)*;
- *continuously reinforced concrete (CRC)*.

Concrete surface slab systems have specific advantages when compared with bituminous pavements; a concrete pavement consists of a system of stiff plates connected together to form a continuous, hinged slab system. The specific advantage of a concrete pavement is that the relatively rigid plates apply load over a wide area. Concrete surface slab systems are useful when:

- low subgrade strengths are anticipated;
- exceptional heavy point loads will be encountered;
- low flashpoint petrochemical spillage may be expected.

The essential features of a concrete slab system are:

- The pavement will include a system of joints which may be natural cracks, as in CRC, or formed joints as in URC or JRC.
- The tensile capacity of the concrete represents a substantial component of the pavement's strength.
- The surface finish must be designed to be resistant to abrasion, frost and also to provide a serviceable skid resistant surface.
- The pavement edge and kerb detail can contribute to the strength of the pavement. An untrafficked shoulder acts as a rigid beam stiffener increasing the pavement strength.

Each of the three distinct forms of concrete pavement are described and discussed in this chapter.

2.2   Unreinforced concrete (URC) pavements

Unreinforced mass concrete pavements consist of a system of (usually) rectangular panels connected together by transverse and longitudinal joints. One of the most important elements of URC pavement design is to ensure that the joints are detailed and, most importantly, spaced correctly. The pavement joints must be arranged to produce a patchwork of roughly square panels with longitudinal joints running in one direction and transverse joints running at 90° – except where geometric constraints absolutely demand that a different angle is used locally. Figure 2.1 shows a typical pavement layout. Joint spacing is controlled by standard practice and is a function of pavement thickness; thicker pavements allow greater joint spacing. Table 2.1 details accepted UK practice for joint spacing. It is noted that the recommended maximum ratio of longitudinal joint spacing to transverse joint spacing is 1.25. Pavement joints may be constructed as plain, dowelled or tied (see Chapter 9); current practice is to construct most highway pavements

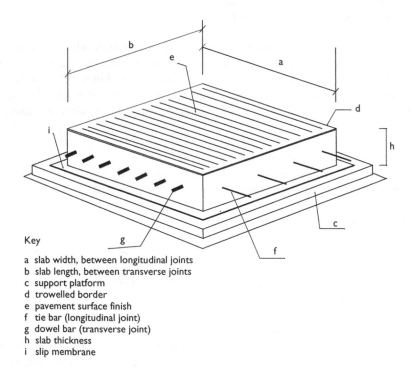

Key
a   slab width, between longitudinal joints
b   slab length, between transverse joints
c   support platform
d   trowelled border
e   pavement surface finish
f   tie bar (longitudinal joint)
g   dowel bar (transverse joint)
h   slab thickness
i   slip membrane

Figure 2.1 A typical URC pavement.

Table 2.1 Maximum UK recommended joint spacing URC pavements [1]

Slab thickness (mm)	Transverse joint spacing		Longitudinal joint spacing	
	Any aggregate (m)	Limestone (m)	Any aggregate (m)	Limestone (m)
150	3.7	4.5	3.7	4.5
200	4.6	5.5	4.6	5.5
250	5.0	6.0	5.0	6.0
300	6.0	7.0	6.0	7.0

with dowels and tie bars. Omitting the steel dowels or tie bars reduces the efficiency of the joint and demands an increase in pavement thickness. The pavement will usually include a slip membrane (see Chapter 10); a bituminous de-bonding membrane is the most efficient form of construction for crack-induced joints, plastic sheet membranes should only be used with dowelled or tied joint details. The maximum recommended highway traffic for untied joints is suggested as 1.5 million standard 80 kN axles (msa – see Chapter 6) or about 1.4 million HGV load applications, although airports and industrial pavements with very thick concrete slabs often make use of plain joints.

Joint sealing is essential for all heavily trafficked pavements. The main function of the joint seal is to:

• Prevent debris from falling down the gap between adjacent slabs. A clogged joint is unable to expand in response to raised temperature. If the pavement becomes restrained, high temperatures can result in 'blow ups' and compression failure.
• Prevent moisture ingress, which can lead to softening of pavement foundation materials and, in extreme cases, 'pumping' of sub-base fines to the surface.

URC applications

Examples of successful forms of mass concrete pavement construction are:

Airfield pavements

Large, strong, 350–450 mm thick, usually slip-formed areas of concrete are constructed with sawn, crack-induced joints. US practice uses dowels and tie bars; UK pavements generally have plain joints. The pavements are built over a slip-formed lean-mix concrete support platform and a bituminous sprayed de-bonding membrane. The lean-mix acts as support to the crack-induced joints, preventing 'faulting'.

A typical UK pavement designed for Boeing 737 loading in accordance with the widely used PSA Design Guide [2] will consist of:

- concrete surface brush finished (see Chapter 10);
- 320 mm URC, 40 MPa 28-day characteristic compressive cube strength, air entrained concrete;
- plain, crack-induced joints;
- a bituminous, sprayed de-bonding membrane;
- 150 mm lean-mix concrete, 20 MPa 28-day characteristic compressive cube strength;
- 5% CBR subgrade.

Bombay city streets

The city currently uses mass concrete as a standard pavement option. The mass concrete pavement is preferred to an alternative bituminous option as the pavement is considered to be more durable in the hot, humid climate.

The Pacific motorway, Queensland, Australia

A substantial section of this recently constructed motorway is built as plain-jointed URC pavement. The highway consists of a section of dual 3-lane and the remainder is dual 2-lane highway. Concrete was used as it was reported [3] to be the most economic solution. The typical pavement construction is as follows:

- 250 mm URC, 40 MPa 28-day characteristic compressive cube strength;
- 125 mm lean-mix concrete;
- 150 mm unbound granular material;
- 300 mm selected capping, minimum soaked or in situ CBR 7%;
- 3% CBR subgrade.

German Autobahn construction

Three different variations on a URC pavement are currently included in the German Federal Standard. Standard URC pavements are constructed with joints at 5 m centres. Each lane is divided into separate slabs. The standard designs are described in Table 2.2. The pavement surface is finished in:

- either a burlap finish in a longitudinal direction (see Chapter 10);
- or a thin layer of 8 mm exposed aggregate (also see Chapter 10).

Table 2.2 German standard URC pavement design options [4]

	Layer thickness (mm)		
URC slab, 40 MPa compressive strength, 350 kg/m^3 cement, 0.38 : 0.44 water–cement ratio, 4% entrained air content	260	270	300
Cement bound base, 15 MPa compressive strength, pre-cracked (see Chapter 3)	150	Membrane over 150	na
Crushed aggregate base	na	na	300
Frost protection blanket	490	480	300

Note
na: not available.

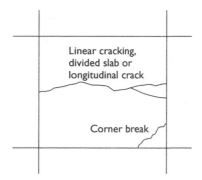

Figure 2.2 Typical structural failure in URC pavements.

URC modes of failure

URC pavements typically fail by some form of tensile fatigue cracking. Further information on pavement failures may be obtained from references [5] and [6]. Repetitive tensile strains within the pavement structure eventually produce pavement surface cracks. The commonest forms of cracking are corner breaks and longitudinal cracks. Figure 2.2 illustrates the typical crack patterns found in a URC pavement. The cracking patterns determine the elements required in a URC pavement design method.

2.3 Jointed reinforced concrete (JRC) pavements

JRC pavements are frequently constructed in a similar manner to URC pavements, in that a conventional longitudinal and transverse joint system is used, but reinforcement is added into the concrete. Figure 2.3 illustrates a typical stylised construction. The reinforcement provides two functions; it controls cracking and holds the joints tightly closed but can also instil additional

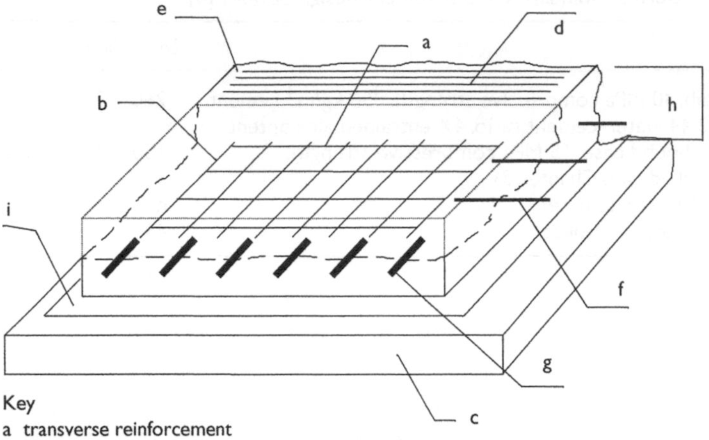

Key
a transverse reinforcement
b longitudinal reinforcement
c support platform
d trowelled border
e pavement surface finish
f tie bar (longitudinal joint)
g dowel bar (transverse joint)
h slab thickness
i de-bonding membrane

Figure 2.3 A typical JRC pavement.

Table 2.3 Maximum joint spacing for JRC pavements [1]

Long slab systems	
Maximum transverse joint spacing	25 m
Maximum longitudinal joint spacing	6 m
Expansion joints	Every third transverse joint
Short slab systems	
Maximum transverse joint spacing	10 m
Maximum longitudinal joint spacing	6 m
Expansion joints	Not required

stiffness into the concrete slab. Reinforced concrete slabs can be constructed as either long or short slab systems (Table 2.3). Long slabs are constructed to a maximum length of around 25 m between transverse joints but typically require expansion joints (see Chapter 9) at each third joint. Short slab systems are constructed up to 10 m long between transverse joints but expansion joints are not usually required. A disadvantage of reinforced slabs is that the wider joint spacing precludes the use of plain joints. The wider transverse joint spacing produces larger thermal movements (as temperature increases and decreases) resulting in increased joint movement when compared with

URC pavements. These movements can only be accommodated using dowel bars or tied joints. The joint opening would be too large for plain joints.

Reinforced systems can be designed as either cracked or uncracked slabs. Reinforcement is typically restricted to a single layer placed at mid-depth in the slab although some designers use separate top and bottom reinforcement. The typical highway standard reinforcement detail is to use 16 mm high tensile bar longitudinal reinforcement of between 500 mm^2 and 1,000 mm^2/m^2 of sectional area with nominal 12 mm diameter bars at 600 mm centres in the transverse direction. A frequently used standard form of industrial hardstanding is to use sheets of A393 (393 mm^2/m length in Grade 460 steel) mesh both at the top and bottom of the slab.

A distinct advantage of fixing the reinforcement in the centre of the slab is that the positive and negative moments are equally balanced thus allowing the slab to flex equally before cracking and failing. The absolute minimum thickness of a reinforced concrete slab designed for trafficking is 150 mm, although 200 mm tends to be a more generally accepted minimum, partly a function of the practical problems involved in providing adequate cover to the reinforcement. Reinforced concrete slabs are generally constructed across plastic de-bonding membranes but bituminous sprayed membranes may also be used. Reinforced concrete slabs are frequently used in place of mass concrete when:

- workmanship and materials are suspect;
- the pavement could be subjected to large unplanned loadings;
- subgrade strength is weak or ill-defined.

JRC applications

Bangkok city streets: short slab JRC system

One of the best examples of the efficient use of reinforced concrete can be found in the city streets of Bangkok. The streets are subjected to a very aggressive trafficking regime in a very hot tropical climate. The city is also subject to subgrade problems that make the effective operation of bituminous pavements problematic. The city is built across a very low-lying river delta, frequently subjecting the streets to severe flooding. Subgrade strengths are exceptionally low; a desiccated surface crust provides most of the strength. Ground water abstraction has also produced massive differential consolidation problems. Historically, poor workmanship has been a further major additional problem, and each of these contributory factors produces an exceptionally difficult environment. But the reinforced concrete pavements operate effectively despite the extreme conditions. Pavements can be seen containing massive cracks, surface polishing and various forms of acid attack, but the system generally functions to an acceptable level of service despite the fact that very little planned maintenance is carried out.

A typical Thai pavement, will carry approximately 10 msa of traffic and consists of:

- concrete surface brush finished;
- 250 mm JRC, 32 MPa 28-day characteristic compressive cube strength;
- longitudinal reinforcement 9 mm diameter at 170 mm centres, transverse reinforcement 9 mm diameter at 300–450 mm centres, all in low-grade plain steel bar;
- bay length 10 m, width one lane, with dowel and tie bar joints;
- plastic sheet de-bonding membrane;
- 150 mm crushed rock sub-base.

Western European industrial sites

Many small industrial sites are built with reinforced concrete slabs if designers lack confidence that a mass concrete alternative would be constructed correctly. If small areas of concrete are required it is frequently preferable to use a reinforced slab and ensure that an unskilled labour force will successfully complete the project without any problems.

A typical 10 msa pavement design consists of:

- concrete surface brushed finish;
- 200 mm JRC, 40 MPa 28-day characteristic compressive cube strength, air-entrained concrete;
- A393 (393 mm^2/m in each direction) Grade 460 reinforcing mesh;
- bay size 10 m by 4–5 m, with dowels and tie bars;
- plastic sheet de-bonding membrane;
- 250 mm, 30% minimum soaked CBR, crushed rock sub-base;
- 5% CBR subgrade.

JRC modes of failure

JRC pavements crack and degrade in a similar manner to URC pavements with the exception that the pavement is more tolerant to cracking. Reinforced pavements are able to maintain a certain level of performance as cracking spreads through the structure. The additional integrity provided by reinforcement prevents a rapid deterioration of the pavement.

JRC pavements frequently suffer from poorly placed reinforcement. Mesh sheets are frequently laid incorrectly supported by bricks or concrete spots resulting in a very poor standard of construction. Proper support chairs are needed to construct a successful pavement.

US experience suggests that long slab systems are less reliable than short slab systems as they are more likely to suffer from mid-panel cracking, joint sealant problems (due to greatly increased opening and closing) and consequent erosion of sub-base.

2.4 Continuously reinforced concrete (CRC) pavements

CRC pavements are an extensively used form of construction employed on large motorway projects. This type of construction is typically economic if large quantities of sand, gravel and water are present on site. If these materials are available, CRC will be substantially cheaper than thick bituminous construction. Several different forms of construction are regularly found, of which the typical construction types are:

Continuously reinforced pavement (CRCP in UK terminology), 100 msa design

- 200 mm CRCP, 40 MPa 28-day characteristic compressive cube strength, surface running slab;
- reinforcement, 0.6% by section area, using 16 mm diameter high tensile, Grade 460 deformed bar and nominal transverse reinforcement in 12 mm bar;
- tied longitudinal joints (assuming the pavement is wider than 5 m);
- bituminous sprayed de-bonding membrane;
- 150 mm lean-mix concrete, 10 MPa 7-day mean compressive cube strength;
- prepared construction platform;
- the pavement ends are anchored into ground beams.

Continuously reinforced roadbase (CRCR in UK terminology), 100 msa design

- 100 mm bituminous surfacing;
- 150 mm CRCR, 40 MPa 28-day characteristic compressive cube strength;
- longitudinal reinforcement, 0.3% by section area, in Grade 460 deformed bar;
- bituminous sprayed de-bonding membrane;
- 150 mm lean-mix concrete 15 MPa 7-day mean compressive cube strength;
- prepared construction platform;
- note: no 'anchorages' are used with CRCR systems.

The CRCP system can, if required, be surfaced with a 35 mm bituminous wearing course; the wearing course is not considered to contribute to the pavement strength.

Some US states use a double mat reinforcement system; the reinforcement mats are placed as top and bottom layers. The two layer system is used to avoid shear failures. The system can produce shear failures, along the plane of maximum shear, across the centre of the slab.

A CRC pavement essentially consists of a regular pattern of cracked square plates connected together by steel reinforcement and aggregate interlock (Figure 2.4). The correct cracking pattern is essential to the efficient operation of the pavement system. Cracking occurs progressively as the concrete shrinks and contracts in response to changes in temperature. The shrinkage cracking has been noted as occurring progressively over the first 100 days of pavement operation. Pavement cracking is controlled by:

- concrete tensile strength;
- concrete shrinkage and the aggregate coefficient of thermal expansion; low coefficients of expansion produce the most favourable cracking pattern;
- the amount and grade of reinforcement;
- concrete temperature at construction; low temperatures produce the best and tightest cracking patterns.

It is accepted that the following criteria are desirable to ensure an efficient pavement system:

- a crack spacing of between 1 and 2 m;
- a maximum crack width of 1 mm;
- a maximum wheel-induced reinforcement stress of 75% of the steel's ultimate strength.

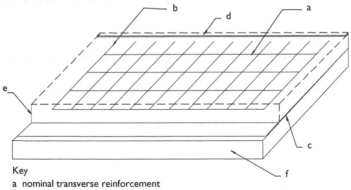

Key
a nominal transverse reinforcement
b longitudinal reinforcement
c bituminous membrane
d 35 mm bituminous surface course (optional)
e slab thickness
f lean-mix concrete sub-base layer

Figure 2.4 A typical CRC construction.

The design of a CRC system is a balancing act; adjusting the percentage reinforcement with the coarse aggregate qualities, cement content and placement temperature to produce the most efficient crack spacing. McCullough and Dossey [7] report that the single most important factor influencing cracking is the aggregate type. Winter placed, low coefficient of expansion, limestone aggregate is noted as producing the most desirable cracking patterns. A recent UK paper [8] has reported a flint aggregate pavement that failed within 5 years of opening. The high coefficient of expansion of the flint aggregate as well as the omission of transverse reinforcement led to a rapid failure of the pavement.

For CRCP, the required quantity of reinforcement is fixed at around 0.6% of the sectional area in most design methods. The Belgian method uses 0.67% [9], the UK 0.6%; US methods fix the rate at 0.65–0.67% [10]. It is noted that increasing the reinforcement percentage beyond these figures can produce poorly performing pavements [9].

A second important detail associated with CRCP construction is the need for ground anchorages at the ends of the CRC length. Two different forms of construction can be employed, a single rigid steel column detail or a system of reinforced concrete ground beams. Each of the anchorage details also includes expansion joints to allow some movement relative to the adjoining pavement construction. If a CRCP pavement is not provided with anchorages and joints, a large bump or ripple will occur at the start of an adjoining bituminous pavement. Chapter 10 contains details of the different anchorage designs.

A thin bituminous 35 mm wearing course is currently considered the most practical UK form of surface finish, overcoming perceived noise and ride quality issues associated with concrete running surfaces. The wearing course is held in place using a bituminous pad coat. The main concrete slab can then be constructed in non air-entrained concrete.

CRC applications

CRC pavements are particularly useful for long continuous pavement construction carrying regular heavy traffic loads.

European motorway projects

There is a long history of Western European motorway and large dual carriageway projects successfully constructed using this form of construction. The most notable recent UK application is a 30 km section of the recently constructed Birmingham Northern Relief Road (M6 Toll) project, which had the following construction:

- 35 mm of 14 mm aggregate, negative texture thin surface course with a 1.5 mm minimum surface texture depth;

- sprayed bituminous emulsion bond coat, 0.8 l/m^2;
- 220 mm CRCP, 40 MPa 28-day cube compressive strength;
- longitudinal reinforcement, 0.6% by section area, using 16 mm diameter Grade 460 deformed bar and nominal transverse reinforcement in 12 mm bar;
- bituminous sprayed de-bonding membrane;
- 200 mm lean-mix concrete, 10 MPa mean 7-day compressive cube strength;
- 5% CBR subgrade;
- pavement ends anchored into ground beams with expansion joints.

A similar Dutch standard design permits the following construction:

- 50 mm negative texture surface course;
- sprayed bituminous emulsion tack coat, 2 layers at 0.3 l/m^2;
- 250 mm CRC, 45 MPa 28-day compressive strength;
- longitudinal reinforcement, 0.7% by section area, using 16 mm diameter high tensile deformed bar and nominal secondary reinforcement in 12 mm bar at 700 mm centres, set at 60° to the longitudinal reinforcement;
- 60 mm asphalt course;
- 250 mm crushed rock or cement bound sub-base;
- pavement ends anchored into ground beams with expansion joints.

Airfield runways

John Lennon Airport runway in Liverpool, UK, was constructed using continuously reinforced concrete techniques and is a notable example of a successfully completed project.

CRC modes of failure

A good summary of current knowledge on modes of distress in CRC pavements can be found in [7] and [10]. The papers describe the following mechanisms:

- punch-outs; resulting from cracking which is too closely spaced;
- crack spalling;
- steel rupture across transverse cracks, resulting in the formation of wide cracks;
- steel corrosion.

Tayabji [10] summarises the following important characteristics for a low maintenance pavement:

- high concrete compressive strength;
- cement treated base layer;
- longitudinal reinforcement of at least 0.59%;
- average crack spacing of 1–1.8 m;
- maximum crack width of 1 mm.

It is accepted that pavement failure occurs progressively as cracking becomes more extensive and crack widths increase until the structural integrity of the pavement is lost. A heavily cracked pavement, with cracks spaced closer than 0.9 m, then becomes susceptible to punch out shear failures.

2.5 Variant slab systems

Fibre reinforced concrete

Steel fibres represent an hybrid alternative to conventional reinforcement. They typically comprise thin strips between 30 and 70 mm in length and are dispersed into the concrete mix during production. Figure 2.5 illustrates typical fibre shapes. Once the concrete layer is placed, the fibres form a continuous reinforcement throughout the slab, the orientation of individual fibres being essentially random. It is not generally economic to include sufficient fibre content for use as CRC – 0.5% by volume is a typical maximum figure – but fibre reinforced concrete slabs are commonly used as JRC in industrial floor slab applications, the benefit being the combination of wide joint spacing (particularly for internal applications where temperature variation is limited) and ease of handling (no reinforcement fixing required).

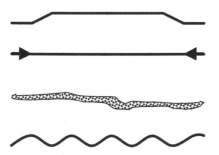

Figure 2.5 Typical shapes used for reinforcing fibres.

Table 2.4 Typical UK pavement designs; 40 MPa strength [1]

Traffic level (msa)	URC (mm)	JRC (mm)	CRCP (mm)
		A393 mesh reinforcement	16 mm Grade 460 0.6%
3% CBR at formation, 250 mm crushed rock sub-base			
1	150	150	200
10	210	190	200
50	270	260	200
100	320	300	210
1,000	na	na	310
3% CBR at formation, 250 mm crushed rock sub-base			
1	150	150	200
10	180	170	200
50	240	230	200
100	270	270	200
1,000	na	na	290

Note
na: not available.

A saving in thickness in comparison with the mass concrete (URC) alternative is also reported by some authors.

2.6 Summary

Each of the different forms of construction can be used for numerous different applications but essentially the following applications are some of the most widely used (Table 2.4):

URC   Airfield aprons, taxiways and runways, heavily trafficked highway pavements in tropical countries;

JRC   Industrial hardstandings, highway pavements in areas of low or uncertain subgrade strength;

CRCP   Motorway pavements on sites with large quantities of site won aggregates.

2.7 References

1. The Highways Agency, *Design Manual for Roads and Bridges*, vol. 7, HD 26/01 'Pavement Design', 2001.
2. Property Services Agency, *A Guide to Airfield Pavement Design and Evaluation*, HMSO, 1989, ISBN 0 86177 127 3.
3. Petrie, E., Heavy duty concrete roads the Australian experience, Britpave Symposium, UK, October 2001.

4. Springenschmid, R., Recent development in design and construction of concrete pavements at German Expressways (Autobahns), 7th International Conference on Concrete Pavements, September 2001.
5. Shahin, M.Y., *Pavement Management for Airports, Roads and Parking Lots*, Chapman and Hall, ISBN 0-412-99201-9.
6. Burks Green and Highways Agency, *Concrete Pavement Maintenance Manual*, The Concrete Society, ISBN 0 946691 89 4.
7. McCullough, B.F. and Dossey, T., Controlling early-age cracking in continuously reinforced concrete pavement, Transport Research Report 1684, Washington DC, January 1999.
8. Cudworth, D.M. and Salahi, R., A case study into the effects of reinforcement and aggregate on the performance of continuously reinforced concrete pavements, 4th European Symposium on Performance of Bituminous and Hydraulic Materials in Pavements, BITMAT 4, Nottingham, UK, April 2002.
9. Verhoevan, K., Cracking and corrosion in continuously reinforced concrete pavements, 5th International Conference on Concrete Pavement Design and Rehabilitation, Purdue University, April 1993.
10. Tayabji, S., Performance of continuously reinforced concrete pavements in the LTPP program, 7th International Conference on Concrete Pavements, September 2001.

Chapter 3

Concrete bases and sub-bases

3.1 Introduction

The foundation beneath a pavement is frequently constructed in either a low strength cracked cement bound material or a higher-quality concrete material. UK practice is to construct heavily trafficked URC, JRC and CRCP pavements across high-quality cement bound materials, the high-quality material providing support to the surface slab system. High-quality cement bound materials can also be used with bituminous materials to form flexible composite pavements. Lower-quality sub-base materials are also used with bituminous pavements but their lack of tensile capacity reduces the structural value of the material.

Flexible composite (or semi-rigid) pavements are distinctly different to surface slab systems in that they obtain their strength from a combination of the structural capacity of the bituminous and cement bound layers. Three different groups of structures can be identified:

- *Roller compacted concrete (RCC) systems*: RCC is principally used in the US and is a strong cement bound material (typically 40 MPa compressive strength) generally used as a complete pavement structure. The material is batched then laid semi-dry using a paver. Cracks are frequently induced as part of the construction process (see Chapter 9). In low-speed applications (ports, industrial areas, estate roads), RCC can actually form the surface; otherwise, a bituminous surface course would be added.
- *Cement bound material (CBM) base systems* (known hereafter as *flexible composite systems*): these are true flexible composite pavements and involve a substantial bituminous bound layer working in association with a structurally significant cement bound layer (typically 10–20 MPa compressive strength). Roller compacted lean-mix concrete is generally used for the cement bound layer, with cracks formed as detailed in Chapter 9.

- *CBM sub-base systems*: these pavements consist of low-strength CBM, with a thick bituminous layer acting as the main structural member in the pavement system. The CBM acts as a granular sub-base.

Figure 3.1 describes a typical sketch comparing the three alternative pavement types.

The choice as to whether a pavement should be considered as true flexible composite or a CBM sub-base system is often not clear cut; the South African standard [1] suggests that it is a function of how the pavement layers are 'balanced'. An alternative explanation would be to suggest that the bituminous and CBM materials will, at the start of the pavement life, carry certain proportions of the load. If the CBM layer is loaded beyond the tensile fatigue capacity of the material, cracking occurs and the CBM layer deteriorates. The cracked layer then begins to act in a similar manner to a conventional crushed rock or granular layer and the bituminous layer carries a higher proportion of the load. It can therefore be concluded that the choice between flexible composite and CBM sub-base is a function of:

- the magnitude of the wheel load;
- thickness of overlying bituminous material;
- CBM material strength and structural capacity;
- environmental conditions;
- the degree of cracking introduced into the CBM layer during the construction process.

A flexible composite pavement works as a bonded system with substantial tensile stress taken by the CBM layer. CBM sub-base systems may begin life as bonded systems but the weaker CBM layer quickly degrades down into a cracked, broken layer similar to a crushed rock.

In all of these pavement types, relatively low-strength CBM can also be used within the pavement foundation, sometimes resulting in two or three consecutive cementitious layers separating the bituminous surfacing from the subgrade.

Figure 3.1 Approximately equivalent composite systems.

3.2   Roller compacted concrete (RCC) systems

The simplest form of construction is the US practice of using RCC pavements. The system is a very simple one in which a semi-dry concrete is laid and compacted in a single layer. The system is summarised in a Portland Cement Association (PCA) paper [2]; essentially it involves the following works:

- A 12% cement content non-air-entrained silica fume OPC blended mix of semi-dry concrete is batched then laid through a paver. A typical highway pavement slab is 225 mm thick.
- The concrete is compacted to give a very high density. Rollers are used to achieve the final closed surface of the RCC layer. A typical ultimate, cored in situ compressive strength will be 40 MPa.
- Joints are frequently formed naturally but in more modern systems sawing or notching is used to create formed joints. The joints are seldom sealed.

The pavement surface can, where required, be finished with a 50 mm layer of Asphalt surface course. Reflective cracks are allowed to form through the Asphalt layer. Industrial sites are frequently left unsurfaced.

RCC applications

The PCA paper [2] notes that the system is extensively used in North America; it is used as an alternative option in many major schemes. The pavements are typically used for:

- freight terminals;
- storage yards;
- minor low-speed rural roads.

The following typical pavements are reported:

- container yards   450 mm RCC, exposed surface
- equipment storage   300 mm RCC, exposed surface
- minor highways   50 mm Asphalt overlay over 225 mm RCC base.

RCC modes of failure

The PCA report [2] fails to describe a manner of failure but it must be assumed that the material will deteriorate in a similar manner to all surface slab systems. The pavements will deteriorate by fatigue cracking which, if bituminous materials are present, will then reflect to the surface.

3.3   Flexible composite systems

Flexible composite pavements are frequently used in place of conventional bituminous construction in situations where substantial quantities of site won aggregates can be commercially exploited, reducing the demand for bitumen and crushed rock aggregate. However, they can be more difficult to construct than conventional systems; the bituminous layers must also be placed and compacted without introducing excessive cracking in the CBM base layer.

Flexible composite applications

A typical UK 20 msa flexible composite motorway pavement design [3] will consist of:

- 190 mm bituminous surfacing;
- 250 mm CBM, crushed rock aggregate, CBM3R well controlled grading, 10 MPa mean 7-day compressive cube strength;
- 150 mm CBM, aggregate and grading much less tightly controlled, CBM2A 10 MPa mean 7-day compressive cube strength;
- CBM layers paver-laid, semi-dry condition, pre-cracked at 3 m centres;
- 250 mm granular or stabilised capping, 15% CBR;
- 5% CBR subgrade.

In the airfield industry, it is common to use multi-layer CBM base beneath a bituminous surfacing. In effect, the combined CBM and bituminous layers are assumed to act together in protecting the underlying subgrade from the often very substantial load from an aircraft undercarriage. This can sometimes demand combined thicknesses of several hundred millimetres. The key decision, upon which not all airport experts agree, is the proportion of these bound layers which should be bituminous and the proportion which may be cement bound. In the UK it is standard practice [4] to insist that at least one-third of the thickness is bituminous, whereas US practice [5] is to permit as little as 125 mm of bituminous (100 mm for light aircraft). Logically, the issue is one of whether reflective cracking (the propagation of CBM cracks/joints through to the surface) is to be tolerated or not. If not, then most climates would probably demand at least 180 mm of bituminous surfacing.

Flexible composite modes of failure

Bonded flexible composite pavements deteriorate by CBM fatigue cracking; as the CBM deteriorates de-bonding occurs. The de-bonded layers produce a weaker pavement. The South African [1] and French [6] design methods contain descriptions of how pavements deteriorate. The standards suggest that immediately after construction the pavement acts as a fully bonded system, the bituminous materials and each layer of CBM is considered to act together producing a single elastic slab response. Initially the CBM layer is assumed to carry most of the load. The first mechanism of deterioration is considered to be fatigue cracking of the CBM layer. Cracking then begins in the de-bonded bituminous surface and subgrade deterioration may also then occur. The stiff uncracked CBM layer is the most structurally significant element in the pavement and acts as the main structural support member; when the CBM layer begins to deteriorate the structural capacity of the system is substantially reduced.

In practice reflective cracking is first noted, since the CBM tends to break into large discrete 'blocks'. Producing stress in the overlying bituminous material.

The South African standard [1] suggests that the long-term structural response of the pavement is a function of the state of the structure within the deterioration process. The final stage of deterioration is observed when the CBM layer acts in a similar manner to a granular sub-base system. South African and French researchers suggest that flexible composite pavements are brittle when compared with bituminous and granular systems.

Figure 3.2 describes a typical section through a flexible composite pavement and describes the point where maximum stress occurs.

3.4   CBM sub-base systems

Cement bound sub-base systems are in many ways similar to flexible composite pavements, the essential difference being that the CBM layer acts in

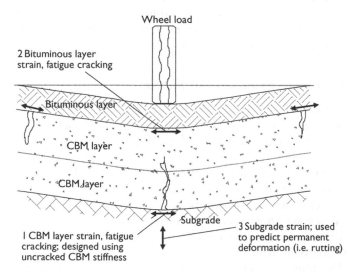

Figure 3.2 Deterioration of a flexible composite pavement.

a manner similar to a granular sub-base. The South African standard [1] suggests that two different classes of CBM sub-base can be produced:

- *Strong systems*, which act as superior granular materials and will have an approximate apparent modulus of 300–500 MPa (compared with 200–300 MPa for most granular bases).
- *Weak systems*, from lower-quality aggregate mixes, producing a modulus of 100–200 MPa.

CBM sub-base system applications

The South African Standard [1] contains a number of examples of cement bound sub-base designs. A typical South African main highway pavement designed to carry 3 msa consists of:

- 30 mm Asphalt surfacing;
- 150 mm CBM Grade C3, 1.5–6.0 MPa compressive strength;
- 200 mm CBM Grade C4, 0.75–1.5 MPa compressive strength;
- 150 mm granular material Grade G7, 15% CBR;
- 150 mm granular material Grade G9, 7% CBR;
- Subgrade, minimum CBR 3%.

The standard also contains extensive lists of other possible flexible composite pavement design options. The French and American standards contain similar examples. The UK experience of CBM sub-base systems essentially exchanges crushed rock for CBM material. The UK sub-base systems have historically made use of CBM with 7-day mean compressive cube strengths of either 4.5 or 7 MPa. New designs taking account of the different possible strengths of pavement foundation resulting from the use of different sub-base materials, are currently in preparation.

CBM sub-base systems modes of failure

This type of pavement effectively behaves as a fully flexible structure and therefore has modes of deterioration typical of such pavements, namely bituminous material fatigue cracking and rutting, together with the possibility of subgrade deformation.

An important factor that is frequently misunderstood is the influence of adverse environmental effects on the durability of CBM materials. CBM sub-base systems can be affected by acid attack, low pH ground water or sulphate attack which may contribute to a lower than expected effective stiffness for the layer. A number of UK examples exist of adverse ground conditions substantially reducing the durability and strength of CBM sub-base systems. This issue would benefit from further investigation. Regrettably no specific published guidance exists linking chemical attack to CBM material specifications.

3.5 Summary

Roller compacted concrete, flexible composite and CBM sub-base systems can offer substantial financial and environmental advantages when compared with more traditional alternatives. The most important fact is that they can be used as direct replacements for conventional crushed rock or bituminous materials, allowing significant reductions in cost and environmental impact. Substantial reductions in the thickness of a bituminous layer can be achieved if correctly designed and constructed CBM layers are used in place of crushed rock sub-base systems. These types of pavement are most economically and successfully used in the following applications:

- *roller compacted concrete* – Industrial hard-standings or low-speed roads in areas with readily available hard aggregates;
- *flexible composite pavements* – high-speed, high-quality highways and airfields, especially when site won aggregates are available;
- CBM *sub-base applications* – any roads where large quantities of low-quality or marginal sand aggregates are available.

3.6 References

1. Committee of Land Transport Officials COLTO, *Structural Design of Flexible Pavements for Interurban and Rural Roads*, Draft TRH4, Pretoria, South Africa, 1996, ISBN 1-86844-218-7.
2. Piggott, R.W., Roller compacted concrete pavements – a study of long term performance, Portland Cement Association, Illinois, RR366, 2000.
3. The Highways Agency, *Design Manual for Roads and Bridges*, vol. 7, HD 26/01 'Pavement Design', 2001.
4. BAA plc, Group technical services aircraft pavements, pavement design guide for heavy aircraft loading, R. Lane, 1993.
5. International Civil Aviation Organisation, *Aerodrome Design Manual*, Part 3 Pavements, 2nd edn, 1983.
6. Laboratoire Central de Ponts et Chaussées, *French Design Manual for Pavement Structures*, Service d'Etudes des Routes et Autoroutes, 1997, D 9511TA 200 FRF.

Chapter 4

Testing and specification

4.1 Introduction

A concrete pavement design method must include a number of elements to successfully complete a pavement design.

- *Index testing* of the bound materials to measure and control strength.
- A *fatigue model* and an accepted set of relationships linking the index test results back to a calculation technique.
- A *standard recognised specification* to classify materials and set acceptance limits to define when materials may be successfully employed within the pavement.

Each of these elements is examined within this chapter.

4.2 Standard strength test methods

The measurement, description, testing and control of the concrete element within a pavement represents a significant design problem. The stiffness and condition of the material within the pavement is difficult to estimate. It is well understood that the strength of concrete increases with time but pavement material strength cannot be directly measured in situ by any of the available test methods. However, a number of different laboratory methods may be employed to test concrete strength (refer to Figure 4.1) and each test is described within this section.

It is noted that none of the tests precisely replicates the exact conditions that occur in the pavement. Pavement material strength may therefore be estimated but never precisely measured.

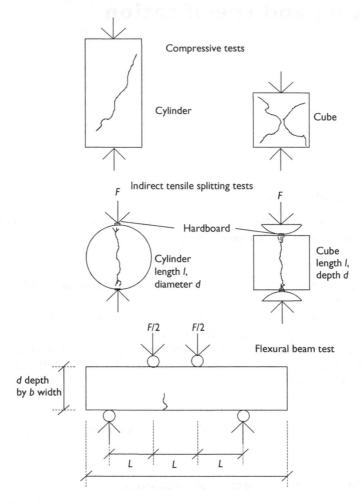

Figure 4.1 Strength testing of concrete samples.

Compressive cylinder (or core) strength

Compressive cylinder tests form the most extensively used method of strength control. The essential features of the test are:

- The size of the sample is controlled; standard cylinders are 150 mm diameter by 300 mm long, although tests may be carried out on any cylinder with a length to diameter ratio between 1 and 2.
- Cylinders may be produced from moulds filled with pavement concrete or cores taken from the completed pavement.

- The method of sampling, compaction and curing is controlled.
- The test is conducted on wet samples.
- The rate of compressive strain application is controlled.
- The rate of reproducibility is around 13%.

The test may be used to measure in situ pavement strength from cores taken out of the pavement. Compressive cylinder strength is accepted as a representative measure of ultimate compressive strength.

Compressive cube strength

The compressive cube strength test forms the basis of UK concrete strength control. Cubes are used in preference to cylinders as they are smaller, easier to store and can be made relatively simply. Two different sizes are regularly used, 100 and 150 mm. Results from both sizes are considered of equal value. Compressive cube strengths are higher than cylinder strengths. When a concrete cube is compression tested the cube's geometry produces a confined testing condition. The steel platens on the compression testing machine restrain the concrete cube at failure and produce higher compression test results when compared to cylinder tests. The essential features of the compressive cube strength test are:

- The method of sampling, compaction and curing is controlled.
- The test samples must be wet when tested.
- The compression-testing machine must have a controlled rate of strain application.
- The 100 mm test is considered to have a 15% reproducibility coefficient.
- The 150 mm test is slightly more precise having a 13% reproducibility.

Compressive cube strength can be related to cylinder strength. CEB-FIP Model Code 90 [2] gives values for equivalent compressive cylinder and cube strengths.

$f_{c,cyl}$	12	20	30	40	50	60 MPa
$f_{c,cube}$	15	25	37	50	60	70 MPa

A well-known empirical relationship is used to link cube and cylinder strength, cylinder strength is approximately equal to 80% of cube strength.

Axial tensile strength

The axial tensile test is a measure of true tensile strength but the test is not generally accepted as a practical method of control in a construction specification. The test is essentially a research tool but will produce the most accurate measure of tensile strength. The test is carried out on standard cylinders; tension is applied from either glued end plates or radial scissor grip side clamps. A preliminary draft European norm, prBS EN 13286-40 [1] exists, based primarily on French experience.

Indirect tensile splitting strength

As an alternative, the indirect tensile (or Brazilian) splitting test is extensively used to assess tensile strength. The test is well known, defined and controlled, but many authorities consider that the test is not a suitable measure of tensile strength. The test is conducted by applying a line load across the diameter of a concrete cylinder (or to the surface of a cube). The essential features of the test are:

- It is usually conducted on cylinders but the UK standard permits the use of cubes.
- Cylinders are loaded through 15 mm by 4 mm hardboard strips, cubes through similar but smaller areas (6 mm square for 150 mm cubes, or 4 mm square for 100 mm cubes).
- The test is conducted using a standard compression-testing machine and the rate of strain application is controlled.
- The test has an 18% reproducibility coefficient.
- The splitting stress is estimated by assuming that failure occurs across the central two-thirds of the failure plane, resulting in the following equation.

Equation 4.1 Tensile splitting strength

$$f_{t,sp} = \frac{2F}{\pi l d} \tag{4.1}$$

$f_{t,sp}$ = tensile splitting stress
  F = applied force
  l = cylinder or cube length
  d = cylinder diameter or cube length

CEB Model Code 90 [2] suggests that a direct relationship exists between ultimate axial tensile strength and tensile splitting strength.

Equation 4.2 Tensile splitting strength to tensile axial strength after [2]

$$f_{t,sp} = 0.9 f_{t,axl} \tag{4.2}$$

Various attempts have been made to relate tensile splitting strength to compressive strength; Oluokun *et al.* [3] presents a relationship.

Equation 4.3 Tensile splitting strength to cylinder strength after Oluokun
et al. [3]

$$f_{t,sp} = 0.214 f_{c,cyl}^{0.69} \qquad (4.3)$$

An important UK paper rejects these hypotheses and strongly asserts that
the tensile splitting test is incorrectly considered to be a measure of tensile
strength. BRE Digest 451 [4] notes that a substantial indirect tensile strength
may be obtained by conducting the test on pre-split cylinders. If a cylinder
is cut along the line of the expected failure plane it is still possible to obtain
50% of the split strength. The Digest notes that the test is a form of shear
test; recording the possible shear strength of a concrete rather than the true
axial tensile strength.

It should be noted that indirect tensile tests can also be used to measure
Young's Modulus, following a similar procedure to that commonly used for
Asphalt [5]. Young's Modulus, though not generally used in specification,
forms an important input to certain pavement design methods. The equation
which applies is as follows and is an uncracked *E* dynamic:

Equation 4.4 Young's Modulus [5] derived from ITSM testing

$$E = \frac{F}{\Delta h \cdot l}(0.273 + v) \qquad (4.4)$$

E = Young's Modulus
Δh = horizontal diametral extension
Other quantities are as defined previously

Flexural strength (or modulus of rupture)

The flexural beam test (see Figure 4.1) is frequently used as a method of
evaluating tensile strength for concrete pavement design and is generally
accepted as the most representative test for assessing the tensile capacity of
concrete pavement materials, since it is considered to establish the extreme
fibre tensile strength in bending. A development of the test is also used to
estimate the ultimate strength and the toughness of fibre reinforced concrete.
The essential features of the test are:

- The depth of the beam (d) is generally equal to the width (b).
- The beam length is between four and five times the depth.
- The ratio of d (or b) to maximum aggregate size should not be less than 3.
- Specimens may be either beams sawn from the pavement or prisms
  manufactured and cured in accordance with the relevant standard.
- Test specimens must be wet.
- The test has a similar level of repeatability to the tensile splitting test
  and has an 18% reproducibility.

- The British Standard test [6] allows compression testing of the discarded beam-ends thus producing a direct relationship between flexural and compressive cube strength.

The equation for flexural strength is as follows:

Equation 4.5 Tensile flexural strength

$$f_{t,fl} = \frac{FL}{bd^2} \tag{4.5}$$

$f_{t,fl}$ = tensile flexural strength (known as 'modulus of rupture' in the USA)
  F = applied load
  L = distance between supporting rollers
  d = beam depth
  b = beam width

Link between axial and tensile strength

It can clearly be seen that the ultimate stress calculated from the beam test is not the true tensile capacity of the concrete. Figure 4.2 illustrates the fact that at failure extreme fibre yield occurs before the sample fails. The axial tensile

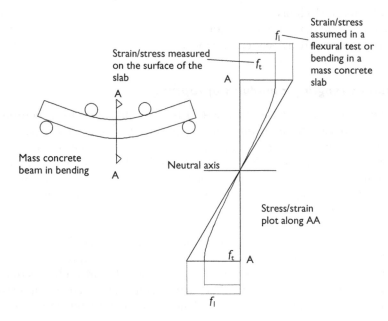

Figure 4.2   Bending stress/strain in a mass concrete beam.

capacity of the concrete therefore turns out to be a function of the test sample size. The ultimate bearing capacity of a mass concrete section is therefore a function of the pavement thickness. This increase in ultimate capacity in a thick concrete pavement is rarely used in design. Most pavement design methods use the allowable ultimate stress as a function of the flexural beam strength. The CEB code [2] suggests a relationship (Equation 4.6) linking axial tensile capacity to a flexural beam strength. The relationship offers an interesting adjustment to the commonly accepted flexural test based fatigue model. The model gives reasonable results on pavement thicknesses between 100 and 250 mm, but will underestimate ultimate strength in pavements thicker than 250 mm.

Equation 4.6 Flexural strength to axial strength relationship after [2]

$$f_{t,axl} = f_{t,fl} \frac{1.5 \left(h_b/0.1\right)^{0.7}}{1 + 1.5 \left(h_b/0.1\right)^{0.7}} \tag{4.6}$$

$f_{t,axl}$ = mean tensile axial strength
f_{fl} = flexural stress at failure
h_b = slab thickness

In the case of 150 mm deep beams, Equation 4.6 reduces to a factor of 0.67 between flexural stress at failure and axial strength. For 100 mm deep beams, this factor becomes 0.6.

Strength for pavement design and construction control

In order to successfully complete the design of a pavement it is essential to understand several key features associated with the estimation of a pavement's true bearing capacity. The key features are, in order of importance:

- the relationship between flexural strength (used in design) and compressive strength (used to control concrete manufacture).
- the estimation (or measurement) of the uncracked Young's Modulus, generally based on compressive strength.
- the fatigue model used within design.

Furthermore the following issues are of key importance in relation to environmental effects on pavement life:

- the thermal expansion coefficient.
- concrete durability.

This section will discuss the issue of concrete strength for pavement design. Sections 4.4–4.7 will address the remaining issues.

Flexural strength to compressive strength relationships

The relationship linking a compressive strength to a flexural strength, and therefore to an axial strength, is not fixed. It is well known and recorded that the link between the expressions is a function of:

- coarse aggregate quality;
- coarse aggregate type (gravel or crushed rock);
- cement type, quantity and quality;
- cleanliness of the fine aggregate.

The following general expression linking compressive cube strength and flexural strength is given in Technical Report 34 [7].

Equation 4.7 General cube strength to flexural strength relationship [7]

$$f_{t,fl} = 0.393 f_{c,cube}{}^{0.66} \qquad (4.7)$$

It is well known that river gravel can give a lower tensile to compressive strength ratio than crushed rock aggregate. Croney and Croney [8] gives two alternative relationships to link flexural strength to compressive cube strength; it is probable the relationships are derived from high-strength, air-entrained concrete.

Equation 4.8 Compressive cube strength to flexural strength after Croney and Croney [8]

$$f_{t,fl} = 0.49 f_{c,cube}{}^{0.55} \quad \text{(for gravels)} \qquad (4.8a)$$

$$f_{t,fl} = 0.36 f_{c,cube}{}^{0.70} \quad \text{(for crushed rock)} \qquad (4.8b)$$

De Larrard [9] quotes similar relationships (Equations 4.9a and 4.9b) for different French aggregate sources. De Larrard's testing was conducted using foil-cured specimens but generated similar relationships to that of Equation 4.7. They are particularly useful in comparing different possible aggregate sources. De Larrard probably used non-air-entrained concrete.

Equation 4.9 Tensile splitting strength to compressive splitting strength after De Larrard [9]

$$f_{t,sp} = k_t f_{c,cyl}{}^{0.57} \qquad (4.9a)$$

$$f_{t,fl} = k_t 1.44 f_{c,cube}{}^{0.57} \qquad (4.9b)$$

Key

Marine flint aggregate	$k_t = 0.442$
Hard limestone aggregate	$k_t = 0.344$
Semi-hard limestone aggregate	$k_t = 0.365$
Basalt aggregate	$k_t = 0.445$
Quartzite aggregate	$k_t = 0.398$

Supporting data

The following data, illustrated in Figure 4.3, was collected by the authors and broadly supports these flexural strength to compressive strength relationships.

Manchester airport, UK

Coarse aggregate, Derbyshire limestone; OPC; 40 MPa 28-day compressive cube mix.

$f_{c,cube}$	28-day mean	57.81 MPa	standard deviation	7.57 MPa
$f_{c,cube}$	7-day mean	47.39 MPa	standard deviation	6.32 MPa
$f_{t,fl}$	28-day mean	5.85 MPa	standard deviation	0.67 MPa

Coventry airport, UK

Coarse aggregate, Leicestershire granodiorite; OPC; 40 MPa 28-day compressive cube mix.

$f_{c,cube}$	28-day mean	58.73 MPa	standard deviation	3.02 MPa
$f_{c,cube}$	7-day mean	47.90 MPa	standard deviation	3.25 MPa
$f_{t,fl}$	28-day mean	6.00 MPa	standard deviation	0.39 MPa

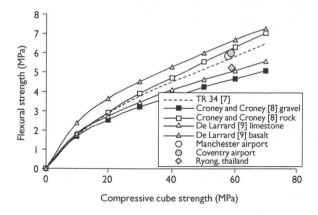

Figure 4.3 Summary of compressive cube to flexural strength data.

General motors site at Ryong, Thailand

Coarse aggregate, granite; non-air-entrained; 40 MPa 28-day compressive cube mix.

$f_{c,cube}$ 28-day mean   58.9 MPa   standard deviation   3.44 MPa
$f_{t,fl}$ 28-day mean   5.26 MPa

The two sets of available UK data suggest that the mean flexural strength of a pavement mix is similar for different mixes produced using a standard specification but the Thai data illustrates that the relationship will change for different concrete mix types.

4.3   In situ pavement compaction and strength

The level of in situ compaction of a concrete will influence the ultimate strength of the pavement; basically the higher the level of compaction, the higher the pavement's ultimate strength. The density and void content of cores taken from the pavement must therefore be monitored and controlled to ensure the concrete (or cement bound base material) will achieve a minimum level of compaction and therefore pavement strength. It should be noted that the core test will measure the total (entrained and other) void content within the concrete and that the greater the volume of entrained air within the concrete the lower the concrete strength will be.

An important element within any design method is the estimate linking index strength tests to the actual achievable strength of the in situ concrete within the pavement, for which very little published data exists. The following approximate relationships can be used to link in situ strength to index testing results.

- 28-day standard tank cured strength = ultimate in situ strength.
- 28-day standard tank cured strength = 1.2 × 28-day in situ strength.

Curing

The precise and correct curing of cement bound materials is an important element contributing to the successful testing of samples. It is noted that all of the recognised testing methods define precise rules for curing concrete samples. The essential features of a curing regime are:

- The samples must be kept for a specific time period within a controlled temperature water bath.
- The test water must be of a controlled quality.
- Tests are typically conducted at 7, 28 and 90 days.

CEB Model Code 90 [2] offers the following relationship to allow the estimation of test strength at key dates after the construction of a pavement.

Equation 4.10 Relationship between strength and time from CEM Code 90 [2]

$$\beta = \exp\left\{S\left(1 - \left[\frac{28}{T}\right]^{0.5}\right)\right\} \tag{4.10}$$

β = factor to adjust 28-day strength
S = cement factor, 0.25 for normal and rapid hardening cement, 0.38 for slow hardening cement.
T = number of days since construction

Equation 4.10 suggests that the 28-day strength can be factored using the following coefficients for OPC mixes.

7-day strength $= 0.78 \times$ 28-day strength
90-day strength $= 1.12 \times$ 28-day strength
360-day strength $= 1.20 \times$ 28-day strength

Young's Modulus for uncracked concrete

The stiffness or Young's Modulus of an uncracked cement bound material cannot be easily directly measured. Standard relationships linking stiffness to strength are accepted as a realistic measure of stiffness. Several researchers have also used ultrasonic techniques to link stiffness density and static modulus. The ultrasonic test is simple to perform and is usually completed on a beam sample. The test will not directly measure stiffness but standard relationships linking ultrasonic speed of transmission to stiffness can be used.

It is therefore usual to derive an approximate Young's Modulus from strength test data. Standard practice is to assume a fixed Poisson's Ratio of 0.15 for static and 0.2 for dynamic pavement calculations and to estimate Young's Modulus based on either compressive cube or cylinder strengths. The level of stress within a concrete pavement layer is usually designed to be relatively low, well below the ultimate strength of the material. It is therefore suggested that dynamic fatigue and strength calculations for pavement design are conducted with Young's Modulus equal to the 'Initial Tangent' value. The Initial Tangent value is the true elastic modulus, free from the effects of creep or yield. CEB Model Code 90 [2] recommends the following for design.

Equation 4.11 Tensile splitting strength to tensile axial strength after [2]

$$E_{ci} = 21.5 C \left(\frac{f_{c,cyl}}{10 \times 10^6} \right)^{0.333} \text{GPA} \qquad (4.11)$$

E_{ci} = Initial Tangent Young's Modulus for dynamically loaded uncracked material
C = correction factor, as follows:
    1.2 for basalt or dense limestone
    1.0 for quartzite
    0.9 for standard limestone
    0.7 for sandstone

The initial creep-free Young's Modulus is not commonly used in structural engineering; most structural engineering calculations use the value adjusted to the 'Secant Modulus', E_c, which the CEB-Code [2] suggests is 0.85 of the Initial Tangent Modulus. Statically loaded pavements such as warehouse flooring should be treated slightly differently. The long-term modulus is significantly different to the short-term value. Specific creep-adjusted values may be estimated but the long-term value may reasonably be taken as two-thirds of the Initial Tangent value.

Equation 4.12 Young's Modulus values [2]

$$E_{ci} = 1.18 E_c = 1.5 E_{\text{long-term}} \qquad (4.12)$$

E_{ci} = dynamic uncracked Young's Modulus (Initial Tangent Modulus)
E_c = static uncracked Young's Modulus (Secant Modulus)
$E_{\text{long-term}}$ = long-term static uncracked Young's Modulus

4.4   Concrete fatigue

A concrete fatigue model is an important element linking together index test results and the structural capacity of the pavement. It is well understood that repeated loads produce tensile stresses which then induce concrete cracking and pavement failure. Pavement life is linked to the magnitude of stress in relation to concrete strength. The difficulty in design is knowing how to link estimated stresses to the number of load cycles which induce failure, and then producing a reliable estimate of pavement life. It is generally accepted that the following factors are directly related to concrete fatigue capacity:

- frequency of loading and amplitude of the loading cycle;
- moisture condition;
- age of the concrete;
- stress condition and stress gradient;
- quality and composition of the concrete.

None of these effects can be precisely calculated or measured.

Basis of fatigue models

The relationship between the number of load repetitions producing a tensile stress, and the flexural strength, is linked using the 'Miner's Law' principle; cumulative damage at different tensile stress levels is summed to produce an estimate of pavement life. A power law is used to relate the effects of stresses at different percentages of the ultimate strength. The power law is also used to link different axle loads back to a 'design axle load' and therefore to an estimated pavement life. Chapter 6 describes this relationship in more detail. Within the context of concrete pavement design, three different types of fatigue model may be found. All of the well-known fatigue models utilise the flexural tensile strength at failure. Tensile splitting strength and axial strength are not generally used. The three main model types involve the following input variables:

1. $f_{t,fl}, \sigma_{t,fl}$
2. $f_{t,fl}, \sigma_{t,fl}, R$
3. $f_{t,fl}, \sigma_{t,fl}, R, t$

$f_{t,fl}$ = flexural tensile strength from the standard beam test
$\sigma_{t,fl}$ = tensile stress calculated under a design load
R = ratio maximum to minimum stress within the loading cycle
t = duration of loading pulse

Within the context of this document discussion is confined to the simplest and most widely used form of model, involving simply flexural strength and tensile stress under load. Further information on concrete fatigue models may be obtained from the paper by Stet and Frenay [10].

Laboratory fatigue data

Numerous researchers have published fatigue data, usually from repeated load flexural beam tests. In general, when the number of load applications

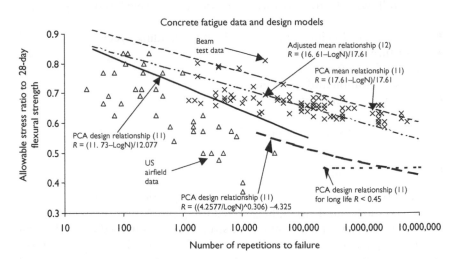

Figure 4.4 Fatigue data and design models.

causing failure is plotted against the ratio of applied flexural tensile stress to flexural tensile strength, the data tends to group within a relatively narrow band. Figure 4.4 includes a set of flexural beam test data. Note that a characteristic of concrete in fatigue is its extreme sensitivity to load level, illustrated by the shallow gradient of the lines in Figure 4.4, and for this reason it is prudent to ensure that a reasonably generous safety margin is designed into a pavement, in order to account for the uncertainties and variations that are inevitably present in such matters as concrete strength and foundation stiffness, not to mention as-constructed slab thickness.

The Portland Cement Association model

The most widely accepted commercial concrete fatigue model is the Portland Cement Association relationship developed by Packard and Tayabji [11]. The Packard model is a semi-empirical relationship linking Westergaard stresses (see Chapter 5) to the 28-day flexural strength from beam tests. Different equations are given for 50% probability and for recommended design, the latter having a greater factor of safety. Darter and Barenberg [12] then proposed a slight adjustment to the relationship for the 50% probability case. It is noted that stress below 0.45 of the ultimate strength is considered to be non-damaging. Equations 4.13 and 4.14 give details of these relationships, which are also illustrated in Figure 4.4. Figure 4.4 also makes the point that the PCA expressions cannot be used to obtain a precise design.

Table 4.1 Summary of coefficients of thermal expansion [13]

Aggregate type	Concrete thermal coefficient $(10^{-6}/°C)$
Flint	20
Quartzite	11.9
Sandstone	11.7
Gravel	10.8
Granite	9.5
Basalt	8.6
Limestone	6.8

Equation 4.13 The PCA [11] fatigue design relationships

$$R = \frac{11.73 - \text{Log}_{10}N}{12.08} \qquad R > 0.55 \qquad (4.13\text{a})$$

$$R = \left(\frac{4.2577}{\text{Log}_{10}N}\right)^{0.306} - 0.4325 \qquad 0.45 < R > 0.55 \qquad (4.13\text{b})$$

$$R \leqslant 0.45 \qquad\qquad N = \infty \qquad (4.13\text{c})$$

Equation 4.14 The PCA [11] 50% probability relationship

$$R = \frac{16.61 - \text{Log}_{10}N}{17.61} \qquad\qquad (4.14)$$

R = the ratio of tensile stress to flexural strength from the 28-day standard beam test
N = the number of load repetitions to failure

Concrete thermal coefficient

The expected temperature gradient through a pavement, leading to differential expansion and contraction and therefore warping, may have a significant effect on the choice of joint design and on the fatigue life of a concrete. The coefficient of thermal expansion is a function of the coarse aggregate type, cement content and humidity. The following coefficients, presented in Table 4.1 as quoted in AASHTO [13] are suitable for pavement design.

4.5   Durability

The ability of a cement bound material to survive its allotted design life without any significant reduction in material quality is an important issue. Surface slabs require a different set of design criteria when compared with cement bound base materials. Surface abrasion, frost resistance and resistance to chemical attack are issues which are of equal importance in the successful construction of a concrete pavement to the more readily understood structural considerations of tensile and compressive strength. The following key issues are important considerations influencing the choice of materials.

Resistance to abrasion and minimum strength requirement

Surface running slabs require a minimum level of strength to avoid erosion from abrasion. The current recommended UK minimum strength (for major highways) is set at a characteristic compressive cube strength at 28-days of 40 MPa. Higher values are suggested in some standards but a 40 MPa mix is the minimum accepted strength threshold to produce abrasion resistance.

Surface slab frost resistance and air-entraining

Frost damage is a major design consideration for concrete slab systems. A very good summary of the current knowledge on the subject is contained within CIRIA Report CP/69 [14]. A number of different mechanisms which produce frost damage occur; they are described as:

- surface scaling;
- production of pop-outs, an aggregate quality problem;
- d-scaling and various forms of internal structural damage.

The basic mechanism of frost damage is generally accepted as a product of the freeze–thaw action of entrapped water within the structure of the concrete matrix. The problem may be prevented by either:

- reducing the size of voids, by adding air entraining agents;
- Producing a high-strength, dense, voidless concrete; the dense material prevents the formation of ice lenses within the structure of the concrete.

The accepted UK and North American method of preventing frost damage is to use air entraining agents. The UK Highways Agency Specification [15] uses 5% ± 1.5% air content for 20 mm aggregate. The entrained air prevents damage by allowing the concrete to strain when subjected to the freeze–thaw cycle. An ongoing debate concerns whether very high cement content concrete needs to be air-entrained. Air-entrained concrete has a reduced strength when compared with an equivalent standard mix, but an increased workability, which then allows the water–cement ratio to be lowered, partially compensating for the loss of strength. TRL investigated the effect of air-entraining agents in LR 363 [16]. Table 4.2 presents the strength data recorded within that report.

Table 4.2 The effect of entrained air on concrete strength [16]

Cement content (kg/m³)	Maximum aggregate size (mm)	Change in flexural strength (%)				Change in compressive strength (%)			
		Air content				Air content			
		3%	4%	5%	6%	3%	4%	5%	6%
310	19	−8	−10	−13	−16	−11	−15	−19	−23
390	19	−6	−8	−11	−14	−10	−16	−21	−27

When the fact that air-entrained concrete has a higher relative volume than non air-entrained material is added into the equation, the economic argument as to whether to entrain air or not becomes even more complex. However, it is the authors' view that producing non-air-entrained concrete in a frost area offers no significant economic advantage.

CBM materials and frost resistance

Cement bound materials constructed within the potential frost penetration zone require a minimum strength and cement content to prevent the occurrence of frost heave or other types of damage. The UK accepted practice is to use a 4% minimum cement content by volume and achieve a minimum compressive cube strength of 4 MPa to avoid frost damage. Weaker cement bound materials may be susceptible to frost.

Pulverised fuel ash (PFA) and air-entrained concrete

PFA and other blast furnace derived ash materials are commonly blended with cement in modern concrete mixes. The furnace ash derived products can offer significant problems when combined with air-entraining agents. TRRL Report 982 [17] examines the viability of using PFA/OPC blended mixes for pavement quality concrete. The report concluded that:

- A 40% cement replaced mix can have a 30% increased flexural strength after 1 year.
- Air content is difficult to control in blended mixes.

The carbon content of PFA has a disrupting effect on air-entraining agents and PFA has a variable carbon content. It is therefore difficult to control the air content in PFA/OPC mixes. The variable nature of the PFA carbon content will produce a mix with an increased variation in compressive cube strength and flexural strength. The technical problems associated with air-entraining an OPC/PFA mix result in some ready-mix companies refusing to produce a blended OPC/PFA mix which complies with the Highways Agency Specification [15].

It is recommended that a concrete mix trial is conducted to confirm that air content is correctly controlled on large pavement projects.

4.6   Specification

Specifications fall into three distinct groups depending on the required function of the layer. The three different groups are:

- *Surface running slabs* This consists of conventional high-quality concrete, are easily recognised in airfields, concrete pavements and hard standings.
- *Strong cement bound materials* This is the group of materials used within flexible composite pavements as the main load carrying layer.
- *Weak cement bound materials* The third group includes low strength replacements for granular layers. The materials are used in the lower pavement structure (sub-base or capping).

Each group of materials has different characteristics and therefore different specifications.

The UK Highways Agency specification [15]

The most comprehensive UK specification is the Highways Agency standard [15]. The relevant sections are found in Series 1000, 800 and 600. These specifications are recommended as the most convenient for use in the UK for pavement works. The standard is used on most UK pavement construction projects; the document is regularly updated and renewed to take account of any changes in knowledge or working methods. Key features are as follows (based on the January 2005 version):

- The standard grade of concrete for surface slabs is 40 MPa 28-day compressive cube strength.
- The UK standard does not define or control tensile strength. The standard is deliberately written to avoid testing flexural strength as it has been found to be impractical within the UK construction industry.
- A minimum cement content of 320 kg/m^3 is also required (where 100% OPC is used); this reduces to an absolute minimum of 255 kg/m^3 within 50 mm of the surface and 220 kg/m^3 elsewhere for blended cements.
- CBM materials for base, sub-base or capping are specified according to their 7-day compressive cube strength and the grades most commonly specified in the past have been 20, 15, 10, 7 and 4.5 MPa. The stronger

Table 4.3 Summary of UK CBM strength assumptions

Material	7-day compressive cube strength (MPa)		28-day strength data (MPa)	
	Mean	Minimum individual	Mean compressive cube	Mean flexural strength
CBMI	4.5	2.5		
CBMIA	10.0	6.5		
CBM2	7.0	4.5		
CBM2A	10.0	6.5		
CBM3G	10.0	6.5	11.0	1.2
CBM3R	10.0	6.5	11.0	1.78
CBM4G	15.0	10.0	16.5	1.82
CBM4R	15.0	10.0	16.5	2.64
CBM5G	20.0	13.0	22.0	2.2
CBM5R	20.0	13.0	22.0	3.2

Note
G = gravel; R = crushed rock.

materials use a more precise control of aggregate grading than the weaker materials.

- CBM grades up to 10 MPa strength may be mixed in place; higher strengths (and more tightly controlled gradings) have to be plant-mixed.
- CBM materials are laid semi-dry and are compacted in a similar manner to granular materials.

Table 4.3 gives a summary of the assumed strength values which have been used in the design of different UK CBM materials. Figures 4.5 and 4.6 illustrate the aggregate grading limits in use prior to the normalisation of European standards.

The UK Ministry of Defence (MoD) specification [18]

This specification [18] is designed for the construction of military airfields; key points in relation to surface slabs are as follows:

- PFA/OPC blended cement mixes are banned.
- Concrete quality is monitored by the estimated in situ compressive cube and flexural strength.
- Three different minimum mean 28 day compressive cube strengths may be used 45, 40 and 35 MPa. The compressive cube strength is then related back to estimated mean in situ cube strengths of 40, 35 and 31 MPa.

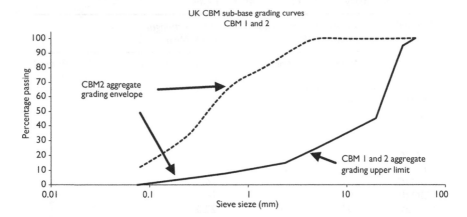

Figure 4.5 UK CBM sub-base grading curves.

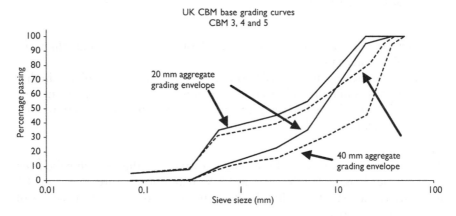

Figure 4.6 UK CBM base grading curves.

- The specification recommends that limestone aggregate is used wherever possible.
- Three alternative mean flexural strengths of 4.5, 4.0 and 3.5 MPa may be used in association with the different cube strengths. The lower strengths are intended for poor quality aggregates where high cement contents may cause alkali–silica reaction.
- The minimum cement content is fixed at 350 kg/m^3.
- The maximum free water–cement ratio is 0.45.
- 20 mm aggregate mixes have a 5% ± 1% entrained air content; 40 mm aggregate mixes have a 6% ± 1% entrained air content.

The MoD specification is difficult to use and cannot be recommended for most general civil engineering works but offers a contrast to the Highways Agency's specification.

The European CBM specification [19]

A complete set of European draft standards now offers the design engineer considerable scope for producing different CBM materials. The standards are described in the following documents:

- prEN 14227-1 to 4 [19] Cement bound mixtures for bases and sub-base;
- prEN 13286-40 to 43 [1] Strength testing methods for CBM materials.

The standards identify 10 different incrementally increasing compressive strength grades of material. The grades vary from 2 to 24 MPa using a characteristic compressive cube test at 28-days. Two additional classes of material are included, namely 'slag bound' and 'fly ash bound' mixtures. The specifications are included in the latest version of the UK Highways Agency specification [15] and are noted in this book for completeness.

UK material control case study

The data presented here, from a UK airfield pavement project, illustrates how a typical 40 MPa characteristic 28-day cube strength mix will vary throughout a contract period. The data covers two cement contents and sources since the initial mix produced declining 7-day cube strengths after the initial approval of the concrete mix. The declining strength brought the material outside the specification requirements. The project's specification was controlled using:

- 7-day compressive cube strength;
- 28-day pavement core strength, converted to an equivalent cube strength;
- 28-day in situ mean flexural strength.

When the material failed the 7-day cube strength requirement, 28-day cube to beam relationships were derived and the estimated in situ flexural strength calculated from the core data. The pavement design required a mean 28-day in situ flexural strength of 4.5 MPa. All of the data presented relates to a typical 40 MPa 28-day characteristic compressive cube strength mix, using limestone aggregate, OPC cement and a 5% air content. The data is summarised in Table 4.4 and in Figures 4.7–4.10.

Table 4.4 Summary of materials testing results from a UK airfield project

	First mix		Second mix	
	Strength [s.d.] (MPa)	*No. of tests*	*Strength [s.d.] (MPa)*	*No. of tests*
Material design				
Mean 28-day compressive cube strength	47.1 [1.0]	4	63.6 [0.9]	4
Mean 28-day flexural strength	6.1 [0.24]	4	7.2 [0.29]	4
(Note: compressive: flexural strength ratios 7.73 and 8.85)				
Site control				
Mean 7-day compressive cube strength	39.3 [6.4]	104	48.6 [4.8]	62
Mean 28-day equivalent compressive cube strength	40.4 [6.6]	43	46.1 [6.2]	43

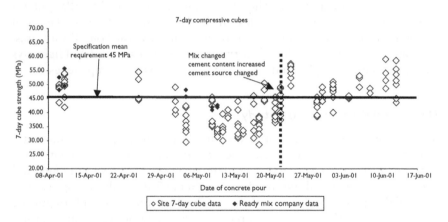

Figure 4.7 Seven-day cube strength data against time.

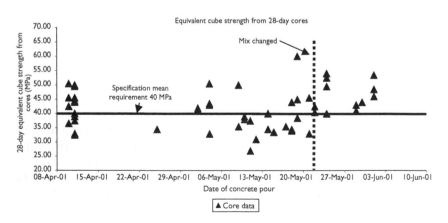

Figure 4.8 Twenty-eight-day core strength data against time.

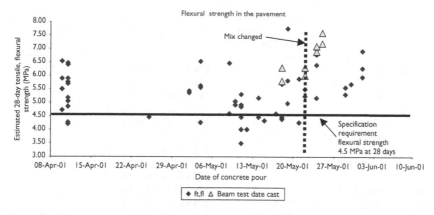

Figure 4.9 Estimated flexural strength at 28 days.

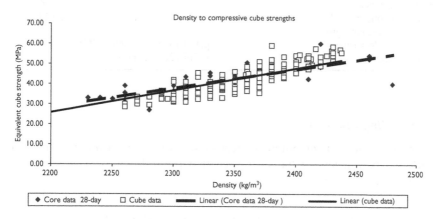

Figure 4.10 Relationship between cube and core density and strength in a pavement project.

The following issues are highlighted by the data:

1 On average, the expected 7-day to 28-day relationship is confirmed.
2 28-day core strength represents a different form of control to 7-day cube or cylinder strength. The additional factors of curing and compaction change the variability of the data. 7-day cubes may not give a reliable estimate of the eventual pavement strength.
3 There is a clear relationship between strength and density, whether from cubes or cores.
4 28-day core-derived cube strength ≈80% of the 28-day laboratory cube strength.
5 The estimated 28-day in situ flexural strength of concrete within a pavement is very variable extending in this case from 3.5 to 7.5 MPa.

Table 4.5 Summary of standard test methods

Characteristic	British Standard BS 1881	European Standard	ASTM
Curing concrete samples	Part 111	EN 12390-2 Part 2	C192
Cylinder manufacture	Part 110	EN 12390-2 Part 2	C470
Compressive testing of cylinders	Part 120	EN 12390-3 Part 3 EN 13286-Part 41, CBMs	C39, C42
Cube manufacture	Part 108	EN 12390-2 Part 2	
Compressive testing of cubes	Part 116	EN 12390-3 Part 3 EN 13286-Part 41, CBMs	
Splitting test of cylinders	Part 117	EN 12390-6 Part 6 EN 13286-Part 42, CBMs	C496
Manufacture of beams	Part 109	EN 12390-2 Part 2	
Flexural strength of beams	Part 118	EN 12390-5 Part 5	C78
Air content	Part 106	EN 12350-7 Part 7	C173, C231
Density	Part 114	EN 12350-7 Part 7	C642
Frost resistance			C666
Seven-day unconfined compressive strength			D1633
Direct tensile strength		prEN 13286-Part 40, CBMs	
Elastic modulus		prEN 13286-Part 43, CBMs	

Concluding remarks

Concretes and cement bound materials are controlled using a number of different parameters; it has been seen that the following can be generally considered as facts:

- Compressive strength is related to tensile strength.
- It is impossible to directly test the true tensile strength of a pavement material.
- A material's durability is an important design consideration; material must be designed for frost resistance and abrasion resistance.
- Standard easily undertaken tests should be used whenever possible in controlling a pavement material. Special complex procedures should be avoided.
- Compaction and void content are important design considerations and should be carefully controlled.
- A standard 28-day compressive strength test, linked both to a tensile strength and to a 7-day compressive strength are the best means of controlling a pavement construction contract.
- Cement type influences pavement strength. Cements with slow strength gain characteristics will produce stronger pavements when compared

with rapid strength gain materials if the pavement is constructed using a 7-day and 28-day compressive strength specification.

- Dynamically and statically loaded pavements will have different stiffnesses. Different calculations should therefore be used for statically loaded pavements when compared with dynamically loaded pavements.
- Table 4.5 summarises some of the most significant British Standards, European Standards and ASTM standards for testing concrete pavement materials.

4.7   References

1. prEN 13286 Parts 40 to 43, Hydraulically bound road materials strength testing methods, 2000.
2. Comité Euro-International du Beton, *CEB FIP Model Code 90*, Thomas Telford, ISBN 0-7277-1696-4.
3. Oluokun, F.A., Burdette, E.G. and Deatherage, J.H., Splitting tensile strength and compressive strength relationship at early ages, *ACI Materials Journal*, 86(2) (1991) 139–144.
4. Clayton, N., *BRE Digest 451 Tension Tests for Concrete*, ISBN 86081 4387.
5. British Standards Institute, Method for determination of indirect tensile stiffness modulus of bituminous mixtures, BS DD213, 1993.
6. British Standards Institute, Testing concrete, BS1881.
7. The Concrete Society, Technical Report 34, 2nd edn, 1994, ISBN 0-946691-49-5.
8. Croney, D. and Croney, P., *The Design and Performance of Road Pavements*, 3rd edn, McGraw-Hill, 1997, ISBN 0-07-014451-6.
9. De Larrard, F., *Concrete Mixture Proportioning*, Spon, ISBN 0-419-23500-0.
10. Stet, M.J.A. and Frenay, J.W., Fatigue properties of plain concrete, 8th International Symposium on Concrete Roads, September 1998, Portugal, Theme 1.
11. Packard, R.G. and Tayabji, S.D., New PCA thickness design procedure for concrete highway and street pavements, Proceedings of the 3rd International Conference on Concrete Pavement Design and Rehabilitation, Purdue, 1985, Section 9, pp. 225–236.
12. Darter, M.I. and Barenberg, E.J., Design of zero maintenance plain jointed concrete pavement, report no. FHWA-RD–77-111, vol. 1, Federal Highway Agency.
13. American Association of State Highway and Transportation Officials, *AASHTO Guide for Design of Pavement Structures*, American Association of State Highway and Transportation Officials, 1992, ISBN 1-56051-055-2.
14. Harrison, T.A., Dewar, J.D. and Brown, B.V., Freeze-thaw resisting concrete its achievement in the UK, CIRIA Report CP 69, funded by the Concrete Society and the Quarry Products Association, 2001.
15. The Highways Agency, *Manual of Contract Documents for Highway Works, Volume 1, Specification*, section 800 and 1000, August 2001, London.

16. Cornelius, D.F., Air-entrained concrete: a survey of factors affecting air content and a study of concrete workability, RRL Report LR 363, Road Research Laboratory, 1970.
17. Franklin, R.E., *The Effects of Pulverised Fuel Ash on the Strength of Pavement Quality Concrete, TRRL Laboratory Report 982*, Transport and Road Research Laboratory, 1981, ISSN 0305-1293.
18. Ministry of Defence, Defence works function standard specification 033, Pavement Quality Concrete for Airfields, London, 1998.
19. prEN 14227 Parts 1 to 4, Unbound and hydraulically bound mixtures – specifications, 2000.

Chapter 5

Concrete slab analysis methods

5.1 Introduction

Three different approaches are commonly used to estimate the structural capacity of concrete slab systems. The methods are:

- *Westergaard* [1–4] *analysis*: semi-empirical, slab stress calculation techniques or finite element calculations.
- *Limit state bearing capacity, plastic design*: methods based around the calculation technique described by Meyerhof [5]. Limit state calculations are again semi-empirical and are rarely employed within any widely used design methods other than for industrial floor slabs [6].
- *Empirical*: calculations based on statistical observational methods derived from evaluations of pavement trials. TRL report RR 87 [7], the basis of the UK Highways Agency pavement designs, is an excellent example.

The background to each approach is examined in more detail within this chapter. It is noted that each of the recognised design methods includes a calibration adjustment which therefore makes each method semi-empirical in nature.

5.2 Westergaard analysis

The Westergaard [2,3] calculation technique is the cornerstone of some of the existing semi-empirical pavement design methods. The system is regrettably frequently misunderstood and is sometimes poorly applied. In some cases standard texts are known to contain typing errors, resulting in incorrect application to pavement design. Ioannides *et al.* [1] reviewed the original papers (2–4) and published a study reappraising each of the commonly applied impressions of the technique. Ioannides *et al.*'s final summary paper is recommended reading for any engineer intending to apply the calculation method. The Westergaard technique should not be considered as a simple

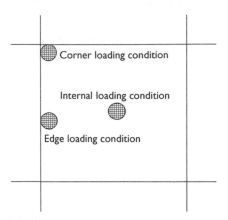

Figure 5.1 Westergaard Internal, Edge and Corner loading.

application of a formula to design. The system requires an understanding of the design assumptions to correctly interpret the calculation. In particular, the deflected shape of the pavement must be understood before the maximum bending strain and therefore stress can be calculated. The system considers three different loading cases, as shown in Figure 5.1.

The following assumptions are made in the calculation:

- The slab is an uncracked laminar.
- The size of the slab is large, such that the internal condition is assumed to be unaffected by the presence of the joints.
- No structural connection is assumed between adjacent slabs in the edge and corner loading cases.
- Slab support is usually modelled by a 'modulus of subgrade reaction' (k), which may be thought of as a series of independent springs, although it is also possible to use an 'equivalent foundation modulus' (E_s).

5.3   Slab bending expressions

The following expressions are used to describe the 'relative stiffness' of a slab system.

Equation 5.1 Radius of relative stiffness; liquid foundation

$$\text{Radius of relative stiffness } \ell = \left(\frac{Eh^3}{12(1 - v^2)k} \right)^{0.25} \tag{5.1}$$

ℓ = radius of relative stiffness (m)

k = modulus of subgrade reaction, = stress/deflection (Pa/m; normally quoted as MPa/m)

E = Young's Modulus of pavement slab (Pa; normally quoted as GPa)

h = thickness of the pavement slab (m)

v = Poisson's Ratio of the pavement slab

Equation 5.2 Radius of relative stiffness; elastic solid foundation

$$\text{Radius of relative stiffness } \ell = \left(\frac{Eh^3(1 - v_s^2)}{6E_s(1 - v^2)} \right)^{1/3} \tag{5.2}$$

ℓ, E, h, v are as defined above

E_s = Young's Modulus of the subgrade, considered as infinitely thick (Pa; normally quoted as MPa)

v_s = Poisson's Ratio of the subgrade

Internal loading condition

The Westergaard calculation is frequently quoted but a precise explanation of the intended deflected slab shape is often neglected. Figure 5.2 is therefore included to identify the key features of an internally loaded slab. They are as follows:

- The radius of relative stiffness, ℓ, is the distance from the centre of the applied point load to the point of contraflexure.
- The point of maximum tensile bending stress occurs on the underside of the slab, under the applied load.
- The point of maximum tensile bending stress on the upper surface of the slab occurs at approximately 2.5ℓ from the applied load.
- The load is a considerable distance from the slab edges.
- The load is uniformly distributed over a circular area.

Equation 5.3 Westergaard [2,3] Internal loading and deflection expressions

$$\text{Internal stress (Pa)} = \left(3P(1 + v)\frac{1}{2\pi h^2} \right)$$

$$\times \left(\log\left(\frac{2\ell}{a}\right) + 0.25 - 0.577215 \right) + \text{BSI2OT} \tag{5.3a}$$

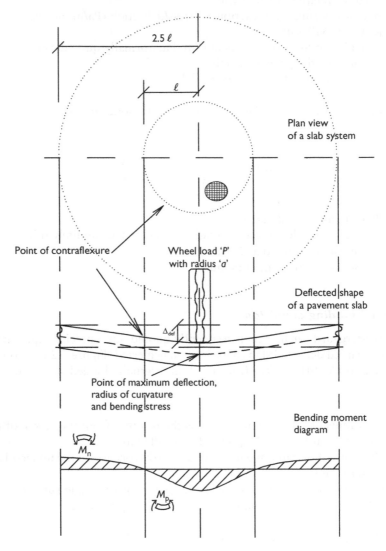

Figure 5.2 The deflected shape of an internally loaded slab.

$$\text{BSI2OT (Pa)} = \left(3P(1+v)\frac{1}{64h^2}\right)\left(\frac{a}{\ell}\right)^2$$

$$\text{Deflection (m)} = \left(\frac{P}{8k\ell^2}\left(1+\frac{1}{2\pi}\right)\right)$$

$$\times \left(\log\left(\frac{a}{2\ell}\right) + 0.577215 - 1.25\right)\left(\frac{a}{\ell}\right)^2 \qquad (5.3b)$$

P = applied load (N)
a = radius of contact area for the point load (m)
v, h, ℓ, k are as defined earlier

Ioannides *et al.* [1] reports that the equations are accurate for the following conditions:

- bending stress; slab larger than 3.5ℓ;
- deflection; slab larger than 8ℓ.

It is noted that the internal loading condition is rarely used in formal design methods.

Edge loading condition

The Edge-loaded case is frequently used in design. The system equates to a wheel running across the Edge of a slab, with zero load transferred to an adjacent slab, and it produces higher stresses and strains than the internal load case. Ioannides *et al.* [1] recommends that the expressions derived by Loseberg [4], based on Westergaard's original equations, compare well with finite element analysis. They are as follows:

Equation 5.4 Loseberg [4] Edge stress and deflection expressions

$$\text{Edge stress (Pa)} = \left(\frac{-6P}{h^2} \right) (1 + 0.5v)$$

$$\times \left(0.489 \log_{10} \left(\frac{a}{\ell} \right) - 0.012 - 0.0063 \left(\frac{a}{\ell} \right) \right) \quad (5.4a)$$

$$\text{Deflection (m)} = \left(\frac{1}{6^{0.5}} \right) (1 + 0.4v) \left(\frac{P}{k\ell^2} \right) \left(1 - 0.760(1 + 0.5v) \left(\frac{a}{\ell} \right) \right)$$

$$(5.4b)$$

where all parameters are as defined previously

The following alternative and frequently presented equation for stress was derived by Kelly [8] based on strains measured in statically loaded pavements:

Equation 5.5 Kelly's [8] empirical Edge stress expression

$$\text{Edge stress (Pa)} = 0.529(1 + 0.54v) \left(\frac{P}{h^2} \right)$$

$$\times \left(4 \log_{10} \left(\frac{\ell}{b} \right) + \log_{10}(39.5)b \right) \quad (5.5)$$

$b = (1.6a^2 + ah^2)^{0.5} - 0.675h$ when $a < 1.724h$

39.5 is added to the original equation to take account of the conversion from metres to inches

Other parameters are as defined previously

However, whichever equation is preferred, the following issues must be considered for sensible pavement design:

- A true Edge-loaded pavement would require a wheel to run along the precise Edge of the pavement slab.
- An adjacent slab or kerb will usually offer some support to the slab, thus reducing both the deflection and the stress.
- The point of maximum tensile bending occurs immediately under the applied load. The point of maximum upper slab surface tensile bending stress occurs approximately 2.5ℓ from the centre of the applied load.

Corner loading condition

The Corner load case forms the basis for many of the well-known mass concrete (URC) pavement design methods. The condition therefore requires understanding if concrete pavement design methods are to be successfully applied. A simplified version of the technique is used in the AASHTO [9] mass concrete pavement design method. The relevant Westergaard expressions are as follows:

Equation 5.6 Westergaard [2,3] Corner stress and deflection expressions

$$\text{Corner stress (Pa)} = \frac{3P}{h^2}\left(1 - \left(\frac{a_1}{\ell}\right)^{0.6}\right) \tag{5.6a}$$

$$\text{Deflection} = \frac{P}{k\ell^2}\left(1.1 - 0.88\left(\frac{a_1}{\ell}\right)\right) \tag{5.6b}$$

a_1 = distance from the load centre to the corner = $2^{0.5}a$

Other parameters are as defined previously

Ioannides *et al.* [1] reported that a finite element analysis would produce bending stresses that are approximately 10% higher than the Westergaard expression. Westergaard [2] states that the point of maximum stress occurs $2(a_1\ell)^{0.5}$ along the Corner angle bisector. Equation 5.10 is an alternative, empirically determined, expression, presented by Pickett [10] and based on field measurements of strain.

Equation 5.7 Pickett's [10] empirical Corner stress expression

$$\text{Corner stress (Pa)} = \frac{4.2P}{h^2}\left(1 - \left(\frac{\sqrt{a/\ell}}{0.925 + 0.22(a/\ell)}\right)\right) \tag{5.7}$$

where all parameters are as defined previously

However, as in the Edge loading case, certain issues reduce the validity of this well-used analysis method, whichever equation is selected.

- A true Corner loading condition will hardly ever occur on a correctly constructed, designed and maintained pavement, since adjacent slabs will usually support the loaded slab.
- Longitudinal joints on highways would normally be located away from wheel paths further reducing the occurrence of Corner load applications.

5.4 Limit state bearing capacity calculation

Limit state or plastic hinge analysis techniques could, in principle, be used to design pavement slabs but the method is not generally applied to external systems; the design technique is generally considered unnecessarily complex for a practical design method. However, limit state analysis techniques are frequently used for the design of industrial floor slabs and are suited to analysis of reinforced pavement systems. The technique may be developed to assess a cracked pavement. The recognised design text is a paper published by Meyerhof [5], which is used to estimate failure loads for internal, Edge and Corner load cases. The recognised equations express failure in terms of a bending moment, M_o, at failure rather than bending stress. The expressions initially appear confusing but, after careful thought, offer an alternative method to describe failure in a concrete pavement system.

Internal loading condition

Under this condition, the tensile bending stress, which occurs immediately below the applied load, is assumed to lead to yield; a crack occurs and a second yield stress is then generated at the upper surface of the pavement, at a distance from the load. The pavement is then assumed to fail, instantaneously, in the form of a number of linked triangular segments (see Figure 5.3), although it should be appreciated that this assumption is really a computational device rather than a precise description of slab failure. Consideration of the geometric shape of the segments, and therefore the rotation induced at each crack, allows the ultimate moment at failure to be determined. Meyerhof [5] also recommends that two extreme load cases should

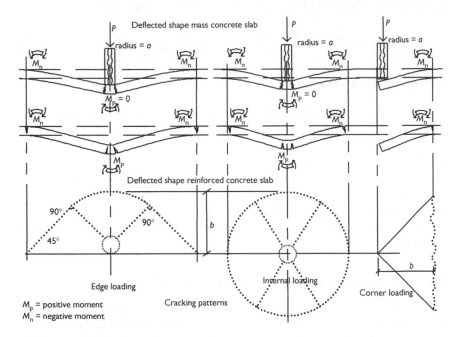

Figure 5.3 The formation of plastic hinges.

be considered: a large slab system, where cracks are at a radius greater than 2ℓ, and a small slab system. The small slab system consists of cracks at radius closer than 2ℓ. Meyerhof's final relationship, which is most frequently used in design, is described as being 'conservative'. For most practical highway load cases it is noted that wheel load contact area is sufficiently small to allow the system to be considered as a point load case. The equations are as follows:

Equation 5.8 Meyerhof [5] Internal moment expression

$$\text{Internal moment (conservative expression) } M_o = \frac{P}{6}\left(\frac{\ell}{\ell + 2a}\right) \qquad (5.8)$$

$$M_o = \text{(large slabs)}\frac{P}{4\pi} = \text{(small slabs)}\frac{P}{2\pi}$$

P = applied load
a = radius of contact area
ℓ = the radius of relative stiffness
M_o = limiting moment of resistance of the slab per unit length

The 'conservative' equations only apply to slabs where the distance between cracks (b in Figure 5.3) is greater than 2.7ℓ and $a/\ell < 0.2$. 'Large slabs' are structures with load redistribution where $b > 3.9\ell$, which corresponds to a situation where joints in the concrete are significantly further apart than 4ℓ. A 'small slab' is a system without load redistribution at failure where the slabs are substantially smaller than 3.9ℓ.

Edge loading condition

The Edge loading condition is the second load case suggested by Meyerhof [5]. The loaded pavement is assumed to produce a tensile cracking pattern in the form of three triangular segments (see Figure 5.3). Tensile cracks appear initially below the wheel load but extend rapidly to lead to the creation of tensile cracks on the upper surface of the pavement. Consideration of the geometry and the degree of rotation at each crack is used to estimate the ultimate bending moment. Meyerhof [5] presents three alternative expressions:

Equation 5.9 Meyerhof [5] Edge moment expression

$$\text{Edge moment (conservative expression) } M_o = \frac{P}{3.5}\left(\frac{\ell}{\ell + 3a}\right) \tag{5.9}$$

$$M_o \text{ (large slabs)} = \frac{P}{2 + \pi}; \; M_o \text{ (small slabs)} = \frac{P}{2 + 0.5\pi}$$

where all parameters are as defined previously

A large slab is a system with joints further apart than 3.9ℓ; a small slab is a structure with joints closer than 3.9ℓ.

Corner loading condition

The Corner loading condition is the most easily understood load case. It assumes that a simple triangular segment of concrete breaks off from a slab through tensile cracking to the upper surface of the pavement. The resulting expressions are as follows:

Equation 5.10 Meyerhof [5] Corner moment expression

$$\text{Corner moment (conservative expression) } M_o = \frac{P}{2}\left(\frac{\ell}{\ell + 4a}\right) \tag{5.10}$$

$$M_o \text{ (large and small slabs)} = \frac{P}{2}$$

For both Edge and Corner loading, the same provisos apply as for the Westergaard calculations; the system will not be precisely replicated in a normally functioning concrete pavement since adjacent slabs will provide support to the pavement, reducing the validity of the calculations.

Limiting moment, M_o

The Meyerhof resisting moment expressions assume that M_o is the sum of the negative and positive resisting moments.

Equation 5.11 Limiting moment of resistance M_o

$$M_o = M_n + M_p \tag{5.11}$$

$M_o =$ limiting moment of resistance per unit length
$M_n =$ negative moment of resistance
$M_p =$ positive moment of resistance

A positive moment creates sagging; a negative moment creates hogging. Positive and negative moments are described graphically in Figure 5.3. The expression can be readily applied to mass concrete pavements; the positive moment, M_p, is not considered to contribute towards resisting failure and is therefore ignored. In the mass concrete case the moment resisting failure, M_o, is taken as equal to M_c; the positive moment M_p is equal to zero and is ignored.

Equation 5.12 Limiting moment of resistance for mass concrete slabs, Internal and Edge load cases

$$M_o = M_n = M_c \tag{5.12}$$

$M_c =$ limiting moment of resistance for mass concrete slab per unit length

The reinforced concrete load case is more complex. The net resultant longitudinal and transverse reinforcement is used to estimate the moment of resistance. It can then be suggested that the following moments of resistance might occur.

Equation 5.13 Limiting moment of resistance for reinforced concrete slabs, Internal and Edge load cases

$$M_o = M_n + M_p \tag{5.13}$$
$$M_n = M_c + M_s$$
$$M_p = M_s \text{ (as } M_c \text{ will } = 0)$$

M_c = limiting moment of resistance for mass concrete slab per unit length
M_s = limiting moment of resistance for reinforced concrete slab per
   unit length

Equation 5.14 Limiting moment of resistance for reinforced concrete slabs,
         Corner load case

$$M_o = M_n + M_p \tag{5.14}$$

$$M_n = M_c + M_s \text{ and}$$

$$M_p = 0$$

In most practical slab analysis methods the calculation is confined to considering the structural contribution obtained from the cracked, reinforced section. The M_c moment is ignored and the slab is considered to have no curvature between the positive and negative cracking regions. The failure moment is considered to be balanced between the concrete compression element and the reinforced steel derived element. Failure calculations would be based around the weakest moment.

Limiting moment, mass concrete slabs

The moment of resistance depends on the tensile capacity of the pavement section when bending occurs around the neutral axis. In the case of a mass concrete (URC) pavement, the tensile capacity can be seen from Figure 5.4. Equation 5.14 applies.

Equation 5.15 Limiting moment of mass concrete slabs

$$M_c = \frac{f_t h^2}{6} \tag{5.15}$$

M_c = limiting moment of resistance per unit length
 f_t = concrete tensile stress at failure
 h = thickness of concrete slab

Limiting moment, reinforced concrete slabs

A reinforced concrete section will have three different limit state conditions:

• Tensile capacity of an uncracked section.
• The reinforcement steel's limiting condition.
• The concrete compression's limiting condition.

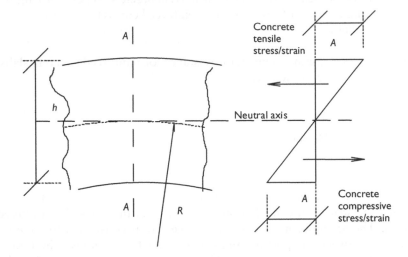

Figure 5.4 Limiting moment in a mass concrete slab.

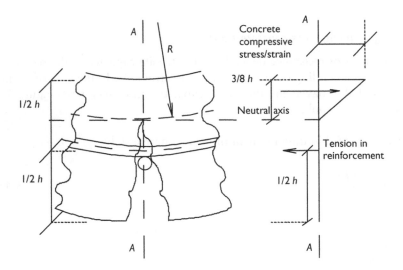

Figure 5.5 Limiting moment in a centrally reinforced concrete slab.

The tensile capacity of an uncracked section is as given in Equation 5.14. In the common case of a centrally reinforced concrete slab the moment resisting failure may be estimated by considering bending around the neutral axis (see Figure 5.5). In many cases, where both longitudinal and transverse reinforcement contribute to resist failure, it is appropriate to use a 'net' moment of resistance as follows:

Equation 5.16 Net moment of longitudinal and transverse reinforcement

$$M_s = (M_t M_l)^{0.5} \tag{5.16}$$

The calculation may be simplified in the centrally reinforced load case to simply identifying the net reinforcement area.

$$a_s = (a_t \, a_l)^{0.5}$$

M_s = net resultant moment a_s = net resultant area of reinforcement
M_t = transverse moment a_t = area of transverse reinforcement
M_l = longitudinal moment a_l = area of longitudinal reinforcement

In the centrally reinforced case, the neutral axis occurs approximately $(3/8)h$ below the compression bending face of the pavement. The limiting moment of the section can then be found by summing the contributions from Equations 5.17 and 5.18.

Equation 5.17 Limiting moment reinforcement, centrally reinforced slab

$$M_S = a_s f_s \left(\frac{h}{8}\right)^2 \tag{5.17}$$

M_S = limiting moment of resistance for centrally reinforced concrete slab per unit length
f_s = steel tensile stress at failure
a_s = net resultant area of reinforcement
h = thickness of concrete

It is noted that the normal rules for reinforced concrete design should be applied to slab design:

- The level of reinforcement should be set to prevent either compressive or tensile failure in the concrete. The slab should be under reinforced.
- The radius of curvature at failure may be estimated using normal bending expressions.

Equation 5.18 Limiting moment concrete compression, centrally reinforced slab

$$M_{cc} = \frac{I_c f_c}{X} \text{general case} = \frac{8 I_c f_c}{3h} \text{for the centrally reinforced case} \tag{5.18}$$

M_{cc} = limiting moment concrete compression per unit length
   d = effective depth of section
   $X = 3/8\ h$ depth as described in Figure 5.5

Equation 5.19 Radius of curvature of the slab

$$R = \frac{EI_{na}}{M} \qquad (5.19)$$

The *I* value around the neutral axis may be approximately estimated using Equation 5.19. The radius of curvature expressions may be used to estimate limiting moments of resistance.

Equation 5.20 *I* neutral axis centrally reinforced slab (describing the section in terms of concrete)

$$I_{NA} = a_s \frac{E_s}{E_c} \left(\frac{h}{8}\right)^2 \qquad (5.20)$$

5.5 Westergaard and Meyerhof techniques compared

The standard equations can be used to estimate the stresses generated from a typical pavement loading case and the results compared. The case selected is as follows:

Slab thickness	0.200 m
Radius of contact area	0.184 m
Point load	64 kN (6.5 tonnes)
Poisson's ratio	0.15
Modulus of subgrade reaction k	40 MPa/m
Concrete elastic modulus	35 GPa
Radius of relative stiffness 'ℓ'	0.879 m

Table 5.1 summarises the resulting stresses for each loading condition. The higher than expected Edge stresses produced by the Loseberg calculation suggests that the assessment may be erroneous. The corner expressions from Westergaard and Meyerhof 'conservative' calculations are seen to produce stresses that are substantially lower than those found in practical test results (i.e. the Pickett equation). Many researchers consider that high Corner stresses are often produced by a near total lack of subgrade support in the Corner region.

The Meyerhof Conservative load case can be used to illustrate the damaging effect of trafficking over or along the edge of a concrete pavement.

Table 5.1 Westergaard and Meyerhof calculated stresses compared

Load case	Stress (MPa)						
	Westergaard				Meyerhof		
	Original [2,3]	Loseberg [4]	Kelley [8]	Pickett [10]	Large slab	Small slab	Conservative
Internal	0.81				0.76	1.52	1.12
Edge		3.67	3.35		1.86	2.68	1.68
Corner	2.48			3.54	4.78	4.78	2.60

If a pavement is edge trafficked, the Conservative expression suggests that a pavement constructed 250 mm thick will give the same ultimate stress as a 200 mm thick internally loaded pavement. In a similar way, by using Kelly's edge loading equation, it can be seen that an unsupported Edge condition demands that the pavement should be 10% thicker than would be the case for a supported condition.

5.6 Treatment of multiple loads

Each of the recognised design methods, Westergaard and Meyerhof, use specific adjustments to the standard relationships to allow the calculation of stresses and strains under multiple wheel loads. In the case of a Meyerhof limit state analysis, it is simply a matter of adjusting the pattern of cracking from the simple case shown in Figure 5.3 to take account of additional loads. It is noted that if loads are spaced further apart than ℓ, multiple point loads will, in general, produce non-critical design load cases. Meyerhof [5] offers the following expressions to assess multiple wheel loads for the internal 'Conservative' design case but notes that the critical design case should be taken as the larger of the single wheel load (Equation 5.11) or either Equations 5.19 or 5.20 as applicable.

Equation 5.21 Meyerhof [5] internal moment treatment for multiple wheels

Dual wheels

$$M_o = \frac{P\ell}{2\pi\ell + 1.8m} \tag{5.21a}$$

Dual tandem wheels

$$M_o = \frac{P\ell}{2\pi\ell + 1.8(m + t)} \tag{5.21b}$$

m = separation of dual wheels
t = separation of tandem wheels
Other parameters are as defined previously

For highway design, the effect of multiple wheel loads can usually be ignored except for the special case of a dual-tyred wheel, where the wheels are so close together that they can be combined and treated as a single load. However, in airfield design, where thick slabs lead to large values of ℓ, this is not the case and it is strongly recommended that the detailed interaction of multiple wheels and joints in the concrete is carefully considered by conducting limit state analyses specific to each design case.

5.7   Factors reducing the validity of semi-empirical design methods

A number of factors reduce the efficiency and effectiveness of the semi-empirical design methods. Many research and design engineers have endeavoured to improve, perfect and complete the quest for an accurate semi-empirical concrete pavement design method but with limited success. The existing design methods, based around the Westergaard and Meyerhof techniques are imperfect. The methods cannot exactly match pavement service conditions. Environmental loading and other non-controlled issues reduce the precision of the design techniques. The following sections describe some of the important limitations to the design equations.

Adjacent slab support

An important issue, which reduces the applicability of the Corner and Edge loading calculations, is the influence of support provided by adjoining pavement slabs. It can be seen that in a correctly designed and constructed pavement the true Corner and Edge loading conditions should never occur. The adjacent slabs will provide support to the loaded system, reducing deflections, rotation and bending stresses. It is generally accepted that a correctly constructed pavement joint will transmit the load across the joint with an efficiency of at least 70%. Figure 5.6 illustrates the structural contribution provided by adjoining slabs.

The Westergaard Corner loading condition produces a stress that is three times larger than the internal loading. The Meyerhof 'conservative' Corner loading position produces a similar result in that the Corner load case produces a stress 2.3 times higher than the internal stress. The Meyerhof 'conservative' Edge loading expression produces a stress 30% higher than that for the internal loading case, which appropriately reflects the influence of zero adjoining slab support. The internal Meyerhof load case might therefore be considered to be an appropriate design case if load transfer between adjacent slabs is maintained which will therefore ensure that the Edge condition should not occur in a well-designed pavement.

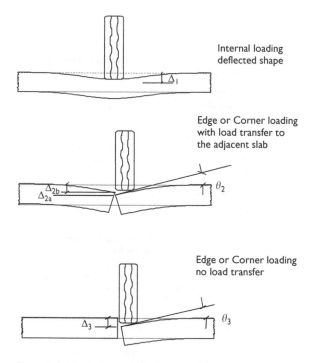

Figure 5.6 The influence of adjoining slabs on pavement deflection.

Note

$\Delta_{2b} > 70\% \times \Delta_{2a}$.

Environmental effects

The second important issue that reduces the applicability of the Westergaard and Meyerhof calculations is the influence of environmentally induced pavement stresses. It can be clearly appreciated that a concrete pavement will be subjected to extensive environmentally induced stresses and strains. Concrete pavements move, warp and distort under the influence of changing temperatures. The following environmental issues, illustrated in Figure 5.7, are noted as producing significant stresses within a pavement. Strong sunlight produces a temperature gradient in the concrete. Expansion of the pavement surface will induce tensile stresses on the underside of the concrete slab. Conversely, night-time conditions induce surface contraction and the occurrence of tensile stresses at the surface.

Temperature changes produce expansion and contraction of the slab. The frictional characteristics of the slab across the subgrade will resist this movement and induce stresses and strains in the pavement.

Differential settlement and the pumping of fines from the sub-base materials will induce 'loss of support' and tensile strain in the concrete pavement. Loss of support is a major reason for pavement cracking in heavily trafficked

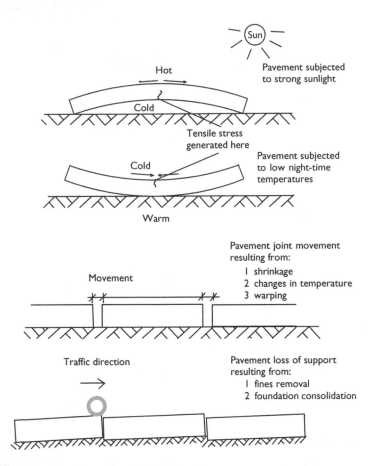

Figure 5.7 Summary of environmental effects.

highways built with unbound sub-bases. To take account of the last point, it is recommended that the Pickett equation is the most suitable for computation of Corner stresses when unbound sub-base is used. With regard to temperature effects, the best-known method for predicting the resulting stresses was produced by Bradbury [11]. His empirically determined equations are given below, together with a graph to determine factors C_x and C_y.

Equation 5.22 Bradbury [11] empirical temperature relationship

Edge warping stress $\sigma_{te} = 0.5\ E\alpha T C_x$ (5.22a)

Internal warping stress $\sigma_{ti} = 0.5\ E\alpha T \left(\dfrac{C_x + C_y}{(1 - v^2)} \right)$ (5.22b)

Constants C_x and C_y depend on the ratio of joint spacing to the radius of relative stiffness and vary from zero at ratios 2 or less to approximately 1 at ratios of over 6.

5.8 Useful relationships for composite slabs

Standard structural design relationships can be applied to composite (that is multi-slab) pavement design. This section of the book presents a number of expressions which designers may find useful. They are presented in terms of the limiting bending moment to be derived from a composite system, but can be applied to Westergaard type calculations by using the 'equivalent slab thickness'.

Two layer unbonded slab system

The following expressions, based around the radius of curvature of a deformed slab, can be used to assess the level of stress within each individual slab layer. The expressions assume that the two slabs act independently.

Equation 5.23 Radius of curvature unbonded two layer slab

$$\text{Radius of curvature } R = \frac{E_1 I_1}{M_1} = \frac{E_2 I_2}{M_2}$$

$$\text{where } I_1 = \frac{h_1{}^3}{12} \quad \text{and} \quad I_2 = \frac{h_2{}^3}{12} \tag{5.23}$$

$h_1, h_2 =$ thicknesses of slabs 1 and 2
$E_1, E_2 =$ Young's Moduli of slabs 1 and 2
$M_1, M_2 =$ bending moments experienced by slabs 1 and 2, per m length

Equation 5.24 Unbonded two layer slab moment of resistance

$$M_0 = M_1 + M_2 = M_1 \left(1 + \frac{E_2 I_2}{E_1 I_1}\right) \text{ or } M_2 \left(1 + \frac{E_1 I_1}{E_2 I_2}\right) \tag{5.24}$$

The moment carried by each slab can then be assessed and the slabs designed to balance, each being at an equal percentage of its ultimate tensile capacity. Equation 5.24 can be used to assess the limiting moment of the system as a whole; the limiting moment is the smaller of the two possible values. It can be seen that balancing the load between the two slabs will produce an optimum design. If either of the slabs is substantially stronger than the other the strongest slab will dominate the ultimate condition. An 'equivalent slab thickness', expressed in terms of the dominant slab (assumed here to be Slab 1) is as follows:

Equation 5.25 Equivalent slab thickness expression

$$h_{eq}^2 = h_1^2 \left(1 + \frac{E_2 I_2}{E_1 I_1}\right) \tag{5.25}$$

Two layer bonded slab system

This load case is significantly more complex than the simple unbonded condition. However, the limiting moment of a bonded system, when $E_1 > E_2$, may be estimated using the following expressions.

Equation 5.26 Distance from base to neutral axis

$$\text{Distance from base to neutral axis; } h_x = \frac{0.5h_1^2 + h_1 h_2 + (E_2/E_1)0.5h_2^2}{h_1 + h_2(E_2/E_1)} \tag{5.26}$$

Equation 5.27 Two layer, bonded system

$$\sigma_{1t} = \frac{1.5M_0}{h_x^2} \quad \text{and} \quad \sigma_{2t} = \sigma_{1t}\frac{(h_1 + h_2 - h_x)E_2}{h_x E_1} \tag{5.27a}$$

Equivalent pavement thickness; $h_{eq} = 2h_x$ $\tag{5.27b}$

$\sigma_{1t}, \sigma_{2t} =$ maximum tensile stresses in Slabs 1 and 2 under applied moment M_0

$M_0 =$ applied moment per m length

Other parameters are as defined previously

Use of these equations will be subject to the following conditions:

- Layers 1 and 2 have to be constructed from similar materials and thicknesses.
- The ultimate shear strength across the joint between the two layers is not exceeded.

Clearly, if reinforcement is present in either or both layers, the calculation becomes more complex still.

Table 5.2 Equivalent concrete thickness from Asphalt surfacing

Concrete slab thickness (mm)	Equivalent thickness of concrete (mm)				
	20 mm Asphalt	40 mm Asphalt	60 mm Asphalt	80 mm Asphalt	100 mm Asphalt
150	2.5	6.1	10.9	16.8	24.0
200	2.4	5.6	9.6	14.5	20.3
250	2.3	5.2	8.8	13.1	18.1
300	2.2	5.0	8.3	12.2	16.7

The effect of a thin Asphalt surfacing

A thin bituminous wearing course is unable to contribute a significant increase in stiffness to the structural system. If the composite slab system is treated as a beam in bending, then the effect of the Asphalt surfacing can be calculated directly by assessing the effect it has on the neutral axis. The shift in the neutral axis means a reduction in the concrete stress for a given bending moment and, to obtain the same stress as without the Asphalt, the concrete slab thickness would have to be reduced. Using this approach, Table 5.2 gives equivalent thicknesses assuming a 3.5 GPa Asphalt surfacing and a 35 GPa concrete slab. However, it has to be admitted that a pavement is not actually directly equivalent to a beam in bending; furthermore, the Asphalt surfacing has a complicating effect on the thermal warping effects due to the efficiency with which it absorbs solar radiation. For this reason it is not uncommon to assume that the surfacing is either negligible or at most equivalent to about one-sixth of its thickness in concrete.

Clearly an assumed contribution of one-sixth of the Asphalt thickness would not be unreasonable based on Table 5.2.

5.9 Implications for pavement design

Despite the admitted differences between the different analysis methods and techniques, the following consistent principles emerge:

- A small increase in concrete slab thickness will produce a substantial increase in pavement stiffness and strength.
- Increasing the thickness of a concrete slab is a more effective method of stiffening a pavement than providing an improved support platform.
- Concrete quality and flexural strength are directly related to pavement performance.
- The magnitude of the applied load is directly related to the resulting pavement stresses, deflections and strains.

- The horizontal dimension of a concrete slab, or the distance between cracks in a cement bound base material, radically alters the potential pavement strength. Large slabs, where cracks are further apart than 3.9ℓ, are substantially stronger than smaller slabs.
- Joint efficiency is very important. Joints that efficiently transmit load from one slab to the next produce substantially more efficient pavements than badly designed joints.
- Corner and Edge loading conditions should be avoided wherever possible. Edge and Corner load cases require special attention in design.

5.10   References

1. Ioannides, A.M., Thompson, M.R. and Barenberg, E.J., Westergaard solutions reconsidered, Transportation Research Record 1043, 1985.
2. Westergaard, H.M., Stresses in concrete pavements computed by theoretical analysis, *Public Roads*, vol. 7, no. 2, April 1926; also Proceedings of the 5th Annual Meeting of the Highway Research Council, Washington DC, 1926, as Computation of stresses in concrete pavements.
3. Westergaard, H.M., Stresses in concrete runways of airports, Proceedings of the 19th Annual Meeting HRB, National Research Council, Washington DC, 1939; also in Stresses in concrete runways of airports, Portland Cement Association, Chicago, Dec 1941.
4. Loseberg, A., *Structurally Reinforced Concrete Pavements*, Doktorsavhandlingar Vid Chalmars Tekniska Hogskola, Gotesborg, Sweden, 1960.
5. Meyerhof, G.G., Load carrying capacity of concrete pavements, Proceedings of the American Society of Civil Engineers, *Journal of the Soil Mechanics and Foundations Division*, June 1962.
6. The Concrete Society, Technical Report 34, 2nd edn, 1994, ISBN 0-946691-49-5.
7. Mayhew, H.C. and Harding, H.M., Research Report 87, *Thickness Design of Concrete Roads*, Transport and Road Research Laboratory, 1987, ISSN 0266 5247.
8. Kelley, E.F., Application of the results of research to the structural design of concrete pavements, Proceedings of the American Concrete Institute, *Journal of the American Concrete Institute*, June 1939, 437–464.
9. American Association of State Highway and Transportation Officials, *AASHTO Guide for Design of Pavement Structures*, American Association of State Highway and Transportation Officials, 1992, ISBN 1-56051-055-2.
10. Pickett, G., A study in the corner region of concrete slabs under large corner loads, *Concrete Pavement Design for Roads and Streets*, Portland Cement Association, Chicago, 1951, pp. 77–87.
11. Bradury, R.D., Reinforced concrete pavements, Wire Reinforcement Institute, Washington DC, 1938.

Chapter 6

Design inputs and assumptions

6.1 Introduction

Chapter 5 has introduced the analysis tools which allow the calculation of stress levels in concrete under load. Earlier, in Chapter 4, the concept of fatigue life was presented, potentially allowing a level of tensile stress to be selected, a certain proportion of the ultimate tensile strength, in order to permit a certain number of stress applications (i.e. wheel loads) to be carried. In theory, therefore, it is now possible to carry out a fully 'analytical' pavement design using 'mechanistic' principles.

Unfortunately, however, the point has been made on numerous occasions that there is no one agreed approach which provides accurate results for all design cases. Furthermore, the inevitable variability in concrete properties on even the best controlled site has been clearly illustrated. Add to this the equally inevitable variability in slab thickness and foundation stiffness, together with uncertainty in future traffic loading, and it is clear that pavement designs should not be based simply on straightforward analysis of idealised cases.

This chapter and the next will therefore address the practical problems associated with real concrete pavement design. This chapter will cover traffic characterisation, pavement foundations and the issue of design reliability; the various practical concrete pavement design methods available will be presented in Chapter 7.

6.2 Traffic characterisation and design life

Power laws and damage factors

The number of load repetitions passing across a pavement is the accepted definition of pavement design traffic. Almost every pavement design method uses some form of calculation to relate real traffic to an equivalent number of standardised loads and several different standardising techniques can be found in different pavement design methods.

Most pavement design methods incorporate a relationship known as a 'power law' to link the magnitude of a real axle load back to a standard design axle. The damage produced by different axle loads is assessed by the power law and the sum of the damage, expressed as a number of standard axles, is then given as the pavement design traffic. The various published power laws use standard engineering fatigue relationships based around Miner's Law [1] to assess the damage from a specific axle load. Equation 6.1 describes the generalised power law relationship used in design.

Equation 6.1 Power law and traffic damage

$$F = \left(\frac{P}{P_s} \right)^n \tag{6.1}$$

F = 'damage factor' applicable to axle load P (also known as 'equivalence factor' or 'wear factor')
P = axle load in kN
P_s = standard axle load in kN
n = relative damage exponent, commonly taken as 4

UK and US standard axle calculations

The most widely used damage calculation approach uses a 80 kN standard axle linked with a fourth power law. This was first derived from trials on a number of different pavement constructions. The 'Law' is therefore empirical, based around observations of pavement trials incorporating both cement bound, unbound and bitumen bound materials. The simple fourth power law approach is used in the UK for all pavement types and thicknesses, and has been used by the Highways Agency (HA) in producing standard average damage factors (known as 'wear factors' in HA documentation) for each class of goods vehicle [2], summing the effects of all the axles. The additional damage due to dynamic effects has also been incorporated by referring to data from weigh-in-motion equipment. Research Report 138 [3] indicates that standard weighbridge recorded axle data should be adjusted by increasing the static axle load damage factor by 1.3 to a give a true 'in motion' damage factor.

The US calculation technique is described in the AASHTO method [4,5] and uses slightly different exponents for:

- bituminous pavements;
- rigid pavements;
- different slab thickness;
- different axle arrangements;
- different pavement terminal service conditions.

Table 6.1 US and UK axle loads and damage factors

Axle load		Damage factors	
Kilopounds (kips)	kN	AASHTO [4] rigid, 250 mm thick n = 4.15	Road Note 29 [6] all pavements n = 4
10	44.5	0.081	0.09
14	62.3	0.338	0.35
18	80.1	1.00	1.00
22	97.9	2.38	2.30
26	115.7	4.85	4.40
30	133.5	8.79	7.60
34	151.3	14.60	12.10
40	178.0	27.90	22.80
50	222.5	69.60	59.84

The method has been calibrated between the following limits:

- pavement thickness, 150–350 mm;
- axle loads up to 180 kN.

Table 6.1 gives a set of typical damage factors appropriate to a 250 mm thick concrete slab comparing the UK, Highways Agency and AASHTO approaches.

Approaches in other counties

Such is the uncertainty regarding the mechanisms of pavement damage and its inevitable complexity that there are a very large number of alternative methods for converting real traffic into numbers for pavement design. For example, a slightly different interpretation of the AASHTO system is used in South Africa [7] where research suggests that damage factors require adjustment for different design cases. The standard uses the same 80 kN (18 kip) standard axle load as AASHTO but recommends that the power law should be adjusted to reflect both pavement type and failure mode. This is based on test results obtained from pavement trials trafficked by the Heavy Vehicle Simulator (HVS) as described by Van Zyle and Freeme [8]. The adjusted exponents are higher for cement bound materials when compared with bitumen bound or granular. The higher exponents reflect the brittle nature of cement bound materials in general. Table 6.2 gives details.

Another interesting development can be found in the French approach to estimating pavement damage. The design manual [9] uses a 130 kN standard

Table 6.2 Damage exponents used in South Africa [7,8]

Pavement type		Exponent 'n'	
Base	Sub-base	Range[a]	Recommended
Granular	Granular	3–6	4
Granular	CBM	2–4	3
CBM (pre-cracked)	Granular	4–10	5
CBM (non–pre-cracked)	Granular	3–6	5
CBM (pre-cracked)	CBM	3–6	4.5
CBM (non–pre-cracked)	CBM	2–5	4.5
Asphalt	CBM	2–5	4

Note
a The higher values usually relate to fatigue failure in the upper pavement, while the lower values relate to rutting.

Table 6.3 Factors used in French method [9]

Pavement type	Traffic (HGV/day)	Power 'n'	Aggression factor	Load type factor		
				Single	Tandem	Tridem
Any	0–25		0.4			
	25–50		0.5			
	50–85		0.7			
	85–150		0.8			
Flexible		5	1	1	0.75	1.1
Asphalt over CBM		12	0.8	1	12	113
URC, JRC		12	1.3	1	12	113
CRC		12	1.3	1	?	?

axle, the heaviest legally allowed on French highways. Several different factors are then used in a complex calculation that essentially reflects the brittle nature of CBM or concrete pavements. Following a power law calculation, the traffic is then adjusted by an 'aggression factor', which depends on traffic intensity, and a 'load type factor', depending on whether single, tandem or tridem axles are involved. Table 6.3 summarises.

The French calculation clearly recognizes that the performance of a concrete pavement is seriously affected by any overloaded axles. The method suggests that overloaded axles must be avoided on CRC, JRC, URC and composite type pavements.

Typical standard axle calculations

Figure 6.1 shows theoretical fully laden trucks permitted in the UK.

As an example, the damage factor applying to the 38 tonne loaded truck is calculated in the following manner (tonnes converted to kN).

- 6 tonne axle damage $(58.9/80)^4$ $= 0.29$ sa.
- 10 tonne axle damage $(98.1/80)^4$ $= 2.26$ sa.
- 7.3 tonne axle damage $(71.6/80)^4 = 0.64 \times 3$ sa.

Total $= 4.47$ sa.

If an exponent of 4.15 was used, as in the AASHTO method, the value would change marginally, to 4.52. Table 6.4 summarises the damage factors for the four truck types shown in Figure 6.1 using the different approaches introduced.

One point is absolutely clear from these computation approaches, especially that from France, and that is that overloading of concrete pavements is a very serious issue. In countries where policing of vehicle load levels is not effective, it is common to find axle loads of 20 tonnes or more and, if the French method is considered appropriate, one 20 tonne axle is equivalent in damaging power to 3,000 normally laden 38 tonne articulated trucks!

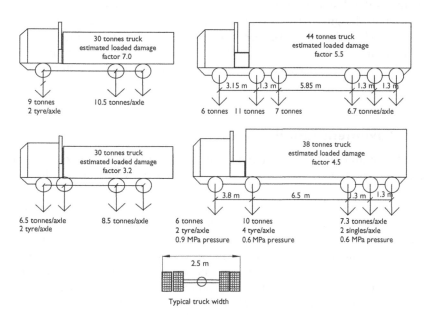

Figure 6.1 Standard UK truck axle loads.

Table 6.4 Summary of damage factors using different methods

Truck type	AASHTO n = 4.15	UK loaded truck n = 4	UK standard practice (2)[a]	South African n = 4.5	French 13 tonne (80 kN) n = 12
30 tonne rigid 3 axle	7.21	6.98	na	7.80	0.17(44.8)
30 tonne rigid 4 axle	3.56	3.17	3.00	3.13	0.01(3.4)
38 tonne articulated	4.52	4.47	3.50	4.58	0.05 (12.4)
44 tonne articulated	5.60	5.51	na	5.84	0.14 (36.8)

Notes
a Based on mean data.
na: not available.

6.3   Aircraft loading

As for highways, there are several different methods in use to describe the traffic load applied to an aircraft pavement. However, the most common technique directly parallels the damage factor approach used for highways. It differs in that there is no 'standard axle', but a particular aircraft type, normally the most damaging common type expected on a given aircraft pavement, is chosen as the 'design aircraft' for that particular pavement. This means that either the Boeing 777 or Macdonald Douglas MD11 usually emerge as the design aircraft for major airfield pavements.

Different design methods, such as the Federal Aviation Authority (FAA) approach [10], then describe techniques for calculating the damaging effect of other aircraft in terms of a number of equivalent design aircraft passes. In some cases the equations relating to rigid pavements are different from those for flexible.

6.4   Pavement foundations

The design and analysis of the pavement foundation is an essential element of pavement design. As introduced in Chapter 5, a measure of foundation stiffness, either a modulus of subgrade reaction (k) or an equivalent foundation modulus (E_s) is required in any analytical approach to pavement design. However, the constructed, in situ stiffness of a pavement foundation is exceptionally difficult to measure and may not be of any significance to the final pavement design since subgrade condition commonly changes after construction of the pavement. Cohesive materials can soften and deteriorate. Frost action may produce a substantially weaker material in periods of spring thaw. Subgrades can become flooded in periods of exceptionally wet weather. The stiffness obtained from an unbound material is also a function of the manner and magnitude of the applied load. And, perhaps of greatest significance, CBM (or other hydraulically bound) layers are expected to deteriorate significantly during a pavement's lifetime. This section therefore discusses the various techniques that may be used to determine an appropriate foundation stiffness for pavement design.

Direct measurement

Three different measurement methods are commonly used to evaluate the quality of a pavement foundation directly, namely:

- *California bearing ratio (CBR)* values (%);
- *modulus of subgrade reaction*, or *k* value (MPa/m);
- *surface stiffness modulus* (MPa).

California bearing ratio (CBR)

The CBR test is the most commonly used method of quantifying subgrade condition but it is *not* a measure of stiffness rather of strength. A CBR value is derived from a standard laboratory or field test when a 49.6 mm diameter steel plunger is pressed into the surface of a material at a constant rate (1 mm/min). The CBR is the percentage of a standard force (derived from a Californian crushed rock) needed to achieve penetrations of either 2.5 or 5 mm. The standard tests are defined in:

laboratory tests, ASTM D1883-94 [11], BS 1377 Part 4 1990 Clause 7 [12]; *field tests*, ASTM D4429-93 [13], BS 1377 Part 9 1990 Clause 4.3 [14].

The Dynamic Cone Penetrometer (DCP) and other forms of cone penetration test can be used as an alternative means of measuring an approximate CBR. Published relationships link DCP values to in situ CBR data; a good summary of these relationships is given in the TRL document Overseas Road Note 31 [15], although it is noted that no such relationship can be precise. The AASHTO standard [16] contains a suggested relationship linking DCP results to the modulus of subgrade reaction k_{762} – see later; it is suggested that this relationship should be used with caution, since the DCP test certainly does not measure stiffness.

Equation 6.2 gives the most commonly used means of interpreting the DCP in the UK, based on use of a 8 kg drop weight and a 60° cone.

Equation 6.2 DCP to CBR relationship [13] Kleyn and Van Heerden
(60° cone)

$$\text{Log}_{10}(\text{CBR}) = 2.632 - 1.28\,\text{Log}_{10}(N = \text{mm/blow}) \qquad (6.2)$$

Modulus of subgrade reaction

The modulus of subgrade reaction (k) is derived directly from an in situ plate-bearing test result. The plate-bearing test is slow and difficult to undertake, and must be used cautiously in the design process. The standard test methods are defined in:

- BS 1377 Part 9 1990 Clause 4.1 [17];
- ASTM D1195-93 [18];
- ASTM D1196-93 [19].

The test descriptions do not specifically identify a method of fixing a k value from the test data.

The test is conducted with a standard 30 in (or 762 mm) diameter plate. It is common practice for pavement design to report the k_{762} value when a plate deflection of 1.25 mm is recorded. This includes any non-recovered deflection. However, ASTM D1195-93 defines a repeated load plate-bearing test. It recommends that six cycles of loading be applied at three different standard plate deflections. The k_{762} value is taken from the ratio of applied stress to recoverable elastic deformation. Although 762 mm is the standard plate diameter, practicalities often dictate that a smaller plate size is used, in which case the Transport Research Laboratory recommend the following adjustment.

Equation 6.3 Adjustment factor to plate size to obtain k_{762}

$$k_{762} = k(1.21\phi + 0.078) \tag{6.3}$$

$k =$ value obtained from plate of diameter ϕ

Surface stiffness modulus

The surface stiffness modulus is the third commonly found parameter describing pavement foundation quality. It is frequently used in association with the FWD (falling weight deflectometer) when it is used to test the surface of a foundation directly. The surface stiffness modulus is effectively a measure of the pavement structural response if the pavement is equated to a single uniform material of infinite depth. The modulus is usually expressed in MPa or GPa and may variously be termed *'stiffness modulus'* or *'modulus'*. It should be used with caution as it is noted that the response obtained from a pavement layer is a function of the magnitude and manner of load application. The Highways Agency [20] give the following equation for determination of surface stiffness modulus from a plate-bearing test (preferably a dynamic plate) result:

Equation 6.4 Surface stiffness modulus M_R

$$\text{Surface stiffness modulus, } M_R = \frac{\pi pr(1 - v^2)}{2y} \qquad (6.4)$$

$p =$ applied stress
$y =$ plate deflection
$r =$ plate radius
$v =$ Poisson's Ratio

Links between foundation parameters

Extreme caution is advised when using data derived for one foundation parameter to obtain equivalent values of another, since the properties described by each parameter are distinctly different. Even k and M_R, though both are measures of stiffness, cannot be exactly related because of the differing inherent assumptions regarding material behaviour. However, a number of published design methods give equations linking the different parameters. It is recommended that each set of expressions is used specifically in association with the particular design method.

The UK design methods derive from work undertaken by the Transport Research Laboratory (TRL) [15,20–23]. The following expressions linking the different parameters are presented in volume 7 of the *Manual for Roads and Bridges* [20]:

Equation 6.5 Surface modulus to CBR relationship

$$\text{Surface stiffness modulus, } M_R = 17.6(\text{CBR})^{0.64} \text{ MPa} \qquad (6.5)$$

(very approximately valid for CBR suited to 2–12%).

Equation 6.6 k_{762} to CBR [23]

$$k_{762} = \left(\frac{\text{CBR}}{6.1 \times 10^{-8}}\right)^{0.577} \times 10^{-3} \text{ MPa/m} \qquad (6.6)$$

The AASHTO design standards [4,16,24] use slightly different relationships to derive expressions linking the various measures. The following relationships are noted in the standards:

Equation 6.7 k_{762} to CBR [5]

Surface stiffness modulus, $M_R = 2.024k_{762}$ MPa   (6.7a)

(for $v = 0.45$)

$$k_{762} = 30.219\text{Ln(CBR)} - 12.4 \text{ MPa/m}$$ (6.7b)

AASHTO (24) suggests that the k_{762} value may vary by the following:

- at 5% CBR the k values can be ±27 MPa/m;
- at 15% CBR the k values can be ±35 MPa/m;
- at 30% CBR the k values can be ±40 MPa/m.

Figure 6.2 illustrates that the UK and US relationships between k and CBR give approximately the same average result, although the AASHTO limits illustrate the uncertainty involved.

Application to design

Whilst direct testing of foundations may be practical and acceptable in some rehabilitation designs, it is not usually possible. The designer therefore has to be able to select an approximate long-term foundation stiffness based on a knowledge of the subgrade and a realistic expectation of the properties to

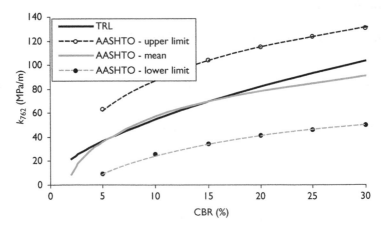

Figure 6.2 UK and US methods linking k_{762} and CBR.

be derived from capping and sub-base layers. Detailed advice on selection of long-term subgrade properties is outside the scope of this work; both AASHTO [4] and Black and Lister [22] present methodologies for assessing the equilibrium conditions likely to apply and, in the case of AASHTO, a technique to take account of variability throughout the year, most notably spring thaw conditions.

The next two sections will present pavement foundation stiffness calculations taken from two different design methods, which allow the designer to determine an appropriate composite pavement foundation stiffness based on properties of individual layers.

The AASHTO [4,24] approach to flexible composite pavement foundations

The AASHTO calculation is undertaken in a similar manner to that for the subgrade, in that it is performed for each individual month of the year and a biased damage-factored average value produced. The original design method [24] also included an adjustment to estimate the 'loss of support' for granular materials, but the Interim Advice Note [16] advises that this adjustment is unnecessary. The calculation therefore follows these steps:

1 An initial estimate is made of the monthly subgrade and sub-base stiffness moduli.
2 The composite k_{762} value on top of the support platform may be obtained by calculating the deflection at time top of foundation level using a multi-layer linear elastic analysis program.
3 These k_{762} values are converted to M_R using Equation 6.7.
4 Damage factors (U_f) are computed for each month of the year using Equation 6.8.

Equation 6.8 Relative damage factor U_f [9]

$$U_f = 1.143 \times 10^3 (M_R)^{-2.32} \tag{6.8}$$

5 The 12 individual values of U_f are averaged and Equation 6.10 followed by Equation 6.7 are then used to derive an equivalent yearly mean k_{762} value.

Equation 6.9 Resilient modulus to relative damage factor [9]

$$M_R = \left(\frac{U_{f\ mean}}{1.143 \times 10^3} \right)^{-0.43} \text{MPa} \tag{6.9}$$

Table 6.5 AASHTO [9] sample effective modulus calculation

Month	Subgrade		300 mm Sub-base stiffness (MPa)	Composite foundation		Relative damage factor U_f (Eq. 6.9)
	CBR (%)	M_R (Eqs 6.7, 6.8) (MPa)		k_{762} (analysis) (MPa/m)	M_R (Eq. 6.7) (MPa)	
1	2	17	100	52	105	0.0233
2	2	17	100	52	105	0.0233
3	3	42	100	99	200	0.0052
4	5	73	150	162	328	0.0017
5	8	102	200	217	439	0.0008
6	8	102	200	217	439	0.0008
7	10	117	250	276	559	0.0005
8	10	117	300	300	607	0.0004
9	8	102	150	201	407	0.0010
10	5	73	100	135	273	0.0025
11	3	42	100	99	200	0.0052
12	2	17	100	52	105	0.0233
Design values				84	171	0.0073

Table 6.5 illustrates a typical calculation for a 300 mm thick crushed rock sub-base laid over a clay subgrade. The calculation shows how a changing subgrade and sub-base strength can be estimated over a yearly cycle and the composite design stiffness calculated for the pavement foundation. Averaging the damage factors in the last column of Table 6.5 gives a U_f value of 0.043. Applying Equation 6.10, gives an equivalent foundation modulus of 80 MPa and Equation 6.7 then gives a design value for k_{762} of 39.5 MPa/m.

UK [23] approach to composite pavement foundations

The UK method [23] uses a similar approach to AASHTO except that the standard UK concrete pavement design method makes use of equivalent foundation modulus (EFM) (surface modulus at top of foundation level) rather than modulus of subgrade reaction k. The approach is based on achieving a similar deflection under load, comparing the real multi-layer case and the equivalent single modulus case, using multi-layer linear elastic computations. Table 6.6 gives a summary of likely foundation design cases, comparing the composite surface stiffness modulus derived from multi-layer computation with EFM tabulated in TRL Report RR87 [21].

A comparison of US and UK values

The following sample calculations (Tables 6.7 and 6.8) are presented as reference cases allowing engineers to compare different calculation

Table 6.6 Summary of RR87 [21] foundation stiffness computations

Subgrade CBR[a]	Pavement foundation materials (mm)				Modulus (MPa)	
	Granular[b] (15% CBR)	Granular[c] (30% CBR)	CBM[d] (10 MPa)	CBM[e] (15 MPa)	Surface stiffness	EFM (RR87)
1.5	600	150	—	—	68	75
2	300	150	—	—	65	72
5	—	225	—	—	89	90
1.5	600	—	150	—	261	278
2	350	—	150	—	268	288
5	150	—	150	—	358	413
15	—	—	150	—	683	649
1.5	600	—	—	150	277	327
2	350	—	—	150	285	311
5	150	—	—	150	383	447
15	—	—	—	150	732	705

Notes
a Subgrade modulus determined from: $E = 17.6 \, CBR^{0.64}$.
b This is 'capping' with an assumed modulus of 70 MPa.
c This is 'Type I sub-base' with an assumed modulus of 150 MPa.
d The '10 MPa' refers to a 7-day cube compressive strength; the assumed modulus is 28 GPa.
e The '15 MPa' refers to a 7-day cube compressive strength; the assumed modulus is 35 GPa.

Table 6.7 Standard support platform cases

Case	Subgrade CBR/k_{762}	First layer details	Second layer details
1	3% 27 MPa/m	250 mm @15% CBR, $E_{max} = 150$ MPa: → EFM = 75 MPa	150 mm @ 30% CBR, $E_{max} = 300$ MPa: → EFM = 150 MPa
2	3% 27 MPa/m	None	250 mm @ 30% CBR, $E_{max} = 300$ MPa: → EFM = 81 MPa
3	5% 37 MPa/m	None	250 mm @ 30% CBR, $E_{max} = 300$ MPa: → EFM = 111 MPa
4	10% 55 MPa/m	None	250 mm @ 30% CBR, $E_{max} = 300$ MPa: → EFM = 150 MPa
5	3% 27 MPa/m	None	150 mm, 10 MPa CBM, $E_{max} = 35$ GPa: → EFM = 5 GPa
6	3% 27 MPa/m	None	200 mm, 10 MPa CBM, $E_{max} = 35$ GPa: → EFM = 5 GPa
7	3% 27 MPa/m	None	320 mm, 10 MPa CBM, $E_{max} = 35$ GPa: → EFM = 5 GPa

Note
E_{ef} = equivalent foundation modulus.

Table 6.8 Standard support platform stiffness calculations

Case	AASHTO [16]		RR87 [21]	762 mm plate test	
	k_{762} (MPa/m)	M_R (MPa)	M_{RR} (MPa)	k_{762} (MPa/m)	M_R (MPa)
1	178	88	125	135	48
2	112	56	78	96	34
3	160	79	111	131	47
4	201	100	140	191	68
5	500	247	381	210	75
6	661	326	489	264	94
7	821	406	608	320	114

techniques. The AASHTO and RR87 assessments apply layer stiffness values in accordance with the recognised published calculation methods.

The EFM values are the effective calculation layer stiffness and are taken as 50% of the maximum potential granular layer stiffness or one-fifth of the uncracked CBM stiffness values. The E_{max} values are the maximum potential layer stiffness which could be generated by either an uncracked CBM or a granular layer.

The M_{RR} value is the surface modulus on top of the support platform when calculated in accordance with RR87. The M_R value is the surface modulus when calculated in accordance with AASHTO. These examples suggest that the different techniques often tend to give similar pavement foundation stiffness values for granular materials. It is noted that this is not the case when considering cement bound materials. The degree of cracking and type of cracking in the cement bound layer leads to confusion. No set specific advice is currently available in any of the design methods to discuss how cement bound sub-base layers should be treated in design. In the absence of specific guidance it is suggested that CBM layers can be safely assessed in design by using an effective layer stiffness of no more than one-seventh of the intact stiffness value, due to the likely presence of cracking. In general, however, since all published design methods are semi-empirical, that is they include a calibration against observed performance (see the Section 'Direct measurement'), the key advice is that each foundation design method should only be used with its associated concrete slab design approach.

6.5 Probability and risk

Whatever design approach is taken and whatever techniques are used to express foundation quality and traffic loading, a key issue is inevitably going to be that of variability and uncertainty. The sources for such variability and uncertainty are numerous and include variations in thickness, concrete strength and joint quality, as well as foundation condition. Pavement design therefore inevitably involves probabilistic principles. Reliability and measured predictability are important issues which must be written into any pavement design method. In general, each pavement design method has to include, either explicitly or implicitly, these two statistical adjustments:

- a calibration adjustment;
- a reliability adjustment.

These two issues are essential to pavement design and will be considered separately. It is noted that the general principles of probabilistic design methods are explained in Appendix EE of the 1986 AASHTO guide [24]. A second good quality paper on the subject is available from Zollinger [25].

Calibration adjustment

A mechanistic pavement design method is usually based around a specific calculation which models a pavement structural response. The structural response is then used to predict pavement failure. Various techniques are used to describe the pavement structural behaviour, typically the Westergaard Corner, Edge and Internal calculations or a statistical interpretation of observed pavement distress. The definition of pavement failure is then adjusted to match the calculations with the observed response seen in pavement trials. The most commonly used method is to:

- use a Westergaard Corner or Internal loading condition and compute maximum tensile stress;
- match the tensile stress back to the PCA fatigue design line [26] to estimate failure.

In this approach, the PCA fatigue design line (see Figure 4.4) already incorporates a calibration adjustment back to real world observed pavements, and so no further calibration is generally required. However, this is not the case for all fatigue models so care is needed to make sure that an appropriate adjustment factor is applied.

In the case of fully empirical designs such as that given in TRL report RR87 [21], the adjustment is performed directly as a statistical interpretation of the observed pavement response. Figure 6.3 illustrates the adjustment, taken from RR87, for a mass concrete (URC) pavement.

The reliability adjustment

The mechanistically based prediction of pavement life is then adjusted to produce a pavement design thickness at an estimated level of reliability,

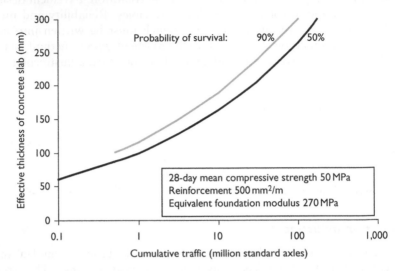

Figure 6.3 RR87 [21] calibration and reliability adjustments for URC.

taking account of the various inevitable uncertainties listed earlier. In simple terms the pavement is thickened to prevent failure. The adjustment can be described in statistical terms as the combined standard deviation of the variables influencing pavement, applied to the design thickness to produce a set level of reliability. The standard deviation is most easily understood in terms of the influence on pavement thickness. When the standard deviation is known, different pavement thicknesses may be estimated corresponding to different levels of reliability. In the case of RR87, Figure 6.3 illustrates the point graphically. In comparison, the AASHTO [4] design method would typically give a 14% increase in slab thickness when moving from a 50% to a 90% design reliability. Equation 6.11 illustrates how design thickness values are typically obtained.

Equation 6.10 Normal distribution reliability adjustment to a design

$$h_{\text{design}} = h_{\text{mean}} + ZS \tag{6.10}$$

$h_{\text{design}} = $ design pavement thickness
$h_{\text{mean}} = $ mean pavement thickness based around a calculation
$Z = $ normal distribution adjustment
$S = $ standard deviation based on the calculation and construction variability

The AASHTO method [4] and its associated reference paper in Appendix EE explains the basis for their statistical adjustment to the basic design equation. The calculation divides the contributory elements associated with the estimation of the standard deviation *S* into four different categories (levels 1–4 follow):

- *Level 0*   the overall sum of the standard deviation, combining Levels 1, 2, 3 and 4;
- *Level 1*   prediction of traffic performance;
- *Level 2*   lack of fit calculation errors involved in the design method;
- *Level 3*   all traffic and design factors;
- *Level 4*   a single traffic and design factor.

The Level 1 and 2 issues are of most interest to the designer. The Level 1 variables are limited to the issues included in the calculation method and are summarised here.

a   slab thickness;
b   material strength;
c   axle load description;
d   drainage and subgrade strength;
e   the definition of failure.

In summary it can be seen that the variation/uncertainty will substantially influence the life of the pavement and therefore confidence in any design. Materials control testing, site workmanship and the construction specification are of high importance to the successful completion of a pavement; a design calculation is just one of the elements required for success.

6.6   Conclusions

This chapter has covered two of the most significant pavement design inputs, namely foundation stiffness and traffic, each of which is also a prime source of uncertainty with regard to eventual pavement performance. This has brought in the subject of probability and risk and the fact that an appreciation of design reliability is an essential element in successful pavement design. The application of design reliability techniques to different pavement types is, of course, complex and varies between calculation methods. It is clear that the following issues, though variously applied, are essentially important to each design method:

1   the accuracy of the calculation method;
2   the construction inputs,

   i   thickness, strength, traffic variation, foundation support;
   ii   the treatment and description of each variable;

3   the required design standard and the expected confidence in design.

It can be seen that the expected confidence in design at different levels cannot easily be checked or estimated. Confidence in design is an abstract concept, which is controlled by construction practice as much as it is by calculation and specification.

6.7   References

1. Miner, M.A., Cumulative damage in fatigue, *Trans ASME 67* (1945).
2. HD 24/96, Highway Agency, *Design Manual for Roads and Bridges*, vol. 7, section 3, HD24/96 Traffic Assessment.
3. Robinson, R.G., Trends in axle loading and their effect on design of road pavements, TRRL Research Report 138, HMSO, 1988.
4. American Association of State Highway and Transportation Officials, *AASHTO Guide for Design of Pavement Structures*, American Association of State Highway and Transportation Officials, 1992, ISBN 1-56051-055-2.
5. American Association of State Highway and Transportation Officials, *AASHTO Guide for Design of Pavement Structures*, vol. 2, August 1986.
6. Road Note 29, *A Guide to the Structural Design of Pavements for New Roads*, 3rd edn, HMSO, 1984, ISBN 0 11 550158 4.
7. Committee of Land Transport Officials, Structural design of flexible pavements for interurban and rural roads, TRH:41996, 1996.
8. Van Zyle, N.J.W. and Freeme, C.R., Determination of relative damage done to roads by heavy vehicles, Annual transportation convention (ATC), 4th CSIR, Pretoria, 6–9 August 1984, vol. C-1.
9. *French Design Manual for Pavement Structures*, Laboratoire Central des Pontes et Chaussées, Service d'Etudes des Routes et Autoroutes, May 1997, D 9511TA 200 FRF.
10. International Civil Aviation Organisation, *Aerodrome Design Manual*, Part 3 Pavements, 2nd edn, 1983.
11. ASTM D1883-94, Standard test methods for CBR (California bearing ratio) of laboratory compacted soils.
12. BS 1377 Part 4 1990 Clause 7, Soils for civil engineering purposes, compaction related tests, pp. 20–26.
13. ASTM D4429-93, Standard test methods for CBR (California bearing ratio) of soils in place.
14. BS 1377 Part 9 1990 Clause 4.3, Soils for civil engineering purposes, in-situ tests, pp. 31–33.
15. Transport Research Laboratory, TRL, Overseas Road Note 31, *A Guide to the Structural Design of Bitumen Surface Roads in Tropical and Subtropical Countries*, 4th edn, London, 1977.
16. American Association of State Highway and Transportation Officials, *Supplement to the AASHTO Guide for Design of Pavement Structures*, American Association of State Highway and Transportation Officials, 1998, ISBN 1-56051-078-1.
17. BS 1377 Part 9 1990 Clause 4.1, Soils for civil engineering purposes, in-situ tests, pp. 26–29.

18. ASTM D1195-93, Standard test methods for repetitive static plate load tests of soils and flexible pavement components for use in evaluation and design of airport and highway pavements, 1993.
19. ASTM D1196-93, Standard test method for non-repetitive static plate tests of soils and flexible pavement components for use in evaluation and design of airport and highway pavements, 1993.
20. *Design Manual for Roads and Bridges*, vol. 7, section 2, HD 25/94 Subgrade Assessment, The Highway Agency, London.
21. Mayhew, H.C. and Harding, H.M., Research Report 87, *Thickness Design of Concrete Roads*, Transport and Road Research Laboratory, 1987, ISSN 0266 5247.
22. Black, W.P.M. and Lister, N.W., Laboratory Report 889, *The Strength of Clay Fill Subgrades: Its Prediction in Relations to Road Performance*, 1979, ISSN 0305-1293.
23. Powell, W.D., Potter, J.F., Mayhew, H.C. and Nunn, M.E., *The Structural Design of Bituminous Roads*, TRRL Laboratory Research Report 1132, 1984, ISSN 0305-1293.
24. American Association of State Highway and Transportation Officials, *AASHTO Guide for Design of Pavement Structures Volume 2*, American Association of State Highway and Transportation Officials, August 1986.
25. Zollinger, D.G., Development of Weibull reliability factors and analysis for calibration of pavement design models using field data, Transport Research Record 1449, 1994, pp. 18–25.
26. Packard, R.G. and Tayabji, S.D., New PCA thickness design procedure for concrete highway and street pavements, Proceedings of the 3rd International Conference on Concrete Pavement Design and Rehabilitation, Purdue, 1985, vol. 9, pp. 225–236.

Concrete pavement design methods

7.1 Introduction

Using the methodologies and techniques introduced in the preceding three chapters, it is now possible to carry out a concrete pavement design from first principles. The key steps are:

1 determine a representative design load case (e.g. a standard axle);
2 work out the number of equivalent design loads during the design life of the pavement;
3 design the pavement foundation to have a certain strength (either k or E_s);
4 select a concrete strength;
5 select an appropriate reliability/calibration factor;
6 combine (2), (4) and (5) and determine an allowable stress;
7 use (1), (3) and (6) in a Westergaard or Meyerhof analysis to design an appropriate slab thickness;
8 ensure that joint spacing/type is compatible with the design assumptions.

This chapter will now present some of the key published concrete pavement design methods, which are based on a combination of theoretical analysis and empirical evidence.

7.2 AASHTO guide for design of pavement structures 1992 [1]

Although scheduled to be superseded soon by a more mechanistic approach, this American classic standard forms the basis of most modern concrete highway pavement design methods. The method uses a simplified version of the Westergaard Corner loading condition to form the basis of the pavement stress calculation. Adjustments are then made to reflect different environmental loadings, pavement configurations or design lives. The document forms the most comprehensive pavement design guide currently available.

The design guide is written in imperial units and is therefore difficult for non-Americans to use, but it is very flexible. It should only be used for major (i.e. heavily trafficked) projects.

The guide will permit the designer to adjust variables as follows:

- A yearly cycle of changes in subgrade strength may be considered allowing frozen subgrades and very dry summer subgrade strengths to be modelled.
- Subgrade drainage conditions and the amount of time the formation is flooded can be considered.
- Cemented or granular sub-bases can be used, at different thicknesses.
- The concrete mix flexural strength may be varied.
- Either dowelled or undowelled joints may be modelled.
- Trafficked or untrafficked shoulders can be considered.
- The loss of support resulting from the pumping of fines from the sub-base may be estimated.
- Different pavement design lives measured in terms of the number of years the pavement is expected to survive can be modelled.
- Different confidence (i.e. reliability) levels may be used.
- The pavement condition at the end of the design life can be varied. The design method describes the condition of the pavement at the end of its life as the terminal serviceability condition.

Concrete thickness design

The principal design equation (Equation 7.1) covers URC, JRC and CRC pavements and is most reliably applied to granular or weak cement bound sub-base materials. The design guide uses mean values for most of the variables used in the estimation of pavement strength. Reliability is expressed in terms of the probability of pavement survival and is used as the design 'safety factor'. The safety factor is applied to the system once.

Equation 7.1 AASHTO [1] basic equation for URC pavement design

$$\text{Log}_{10} W_{18} = Z_R S_o + 7.35 Q - 0.06 + V + H_a \left(\text{Log}_{10} \left(\frac{A}{B(N-F)} \right) \right)$$

(7.1)

$$Q = \text{Log}_{10}(D+1)$$

(7.1a)

$$V = \text{Log}_{10} \left(\left(\frac{\Delta \text{PSI}/(4.5-1)}{(1 + (1.624 \times 10^7/(D+1)^{8.46}))} \right) + 1 \right)$$

(7.1b)

$$H_a = 4.22 - 0.32 P_t$$

(7.1c)

$$N = D^{0.75}$$

(7.1d)

$$A = S'_c C_d (D^{0.75} - 1.132) \tag{7.1e}$$

$$B = 215.63J \tag{7.1f}$$

$$F = 18.42 \Big/ \left((E_c/k)^{0.25} \right) \tag{7.1g}$$

$$\Delta PSI = P_o - P_t - \text{weather factor} \tag{7.1h}$$

where

W_{18} = traffic loading in standard 18 kip (80 kN) axles
D = pavement thickness in inches
E_c = Young's Modulus of concrete in psi
k = modulus of subgrade reaction in pci
S'_c = the mean 28-day modulus of rupture – that is, flexural strength (taken as 863 psi, 5.9 MPa for UK 40 MPa concrete)
S_o = the standard deviation of the data used to construct the pavement (taken as 0.29)
Z_R values (for a statistically normal distribution):   95%: −1.65; 85%: −1.04; 75%: −0.68; 50%: 0
J values: $J = 3.2$ trafficked edge and shoulders with dowel bar joints
             $J = 4.1$ trafficked edge and shoulders with no dowel bars in joints
             $J = 2.8$ untrafficked edge with dowel bar joints
             $J = 3.9$ untrafficked edge with no dowel bar joints
Drainage coefficients: $C_d = 0.75$ poorly drained subgrade, wet for 25% of the year
                        $C_d = 1.0$ fairly well drained subgrade, wet from 1% to 5% of the year
                        $C_d = 1.25$ excellent subgrade drainage system, wet for <1% of the year
ΔPSI = change in pavement serviceability index during pavement life
P_o = initial serviceability condition (taken as 4.5 in the AASHRO road test)
P_t = terminal serviceability condition (taken as 2.5 for major high-quality highways and 2.0 for minor projects where a poorer ride quality is permitted)

The 'weather factor' is specific to the intended climatic zone, drainage conditions and subgrade type. The example in the AASHTO standard takes 0.85 for frost susceptible subgrades and 0.35 for non-frost susceptible subgrades. Thus the following values of ΔPSI may be used for a 20-year design.

Table 7.1 Summary of AASHTO design example

Traffic (msa)	Slab thickness (mm)			
	Untrafficked edge dowelled	Trafficked edge dowelled	Untrafficked edge undowelled	Trafficked edge undowelled
	$J = 2.8$ m	$J = 3.2$ m	$J = 3.9$ m	$J = 4.1$ m
0.5	0.120	0.131	0.150	0.155
1	0.137	0.151	0.172	0.178
5	0.188	0.203	0.228	0.234
10	0.211	0.228	0.254	0.260
50	0.273	0.293	0.324	0.333
100	0.303	0.325	0.359	0.368
500	0.385	0.412	0.455	0.466

High-quality roads, frost sensitive sub-base
$$= 1.15\,[4.5 - 2.5 - 0.85]$$
Most UK high-quality roads, non-frost sensitive sub-base
$$= 1.65\,[4.5 - 2.5 - 0.35]$$
Low-quality road to failure, frost sensitive sub-base
$$= 1.65\,[4.5 - 2.0 - 0.85]$$
Low-quality road to failure, non-frost sensitive sub-base
$$= 2.15\,[4.5 - 2.0 - 0.35]$$

A nomograph is contained within the standard which may be used as an alternative to Equation 7.1.

Table 7.1 gives standard designs derived from Equation 7.1 for the following design parameters:

E_c = concrete stiffness modulus, taken as 35 GPa

S'_c = 28-day modulus of rupture, taken as 5.9 MPa (863 psi) for 40 MPa concrete

S_0 = standard deviation of the data used in the pavement assessment, taken as 0.29

J = 2.8 for tied shoulders

k = 40 MPa/m (149 pci), appropriate to the crushed rock sub-base option

C_D = drainage coefficient, taken as 1 for UK conditions

ΔPSI = 1.65 (P_0 taken as 4.5, P_t = 2.5, weather factor = 0.35)

Z = 90%

Reinforcement design for JRC pavements

The additional strength provided by light reinforcement is noted as making a negligible contribution towards the overall pavement strength.

The standard pavement thicknesses for URC (Equation 7.1) can therefore also be applied to lightly reinforced concrete pavement sections. The AASHTO guide proceeds to explain that the purpose of the reinforcement is to keep any cracks which develop tightly closed.

The reinforcement is designed separately for longitudinal and transverse directions. The method uses a simple calculation (Equation 7.2), balancing the percentage reinforcement to the ultimate tensile strength of the steel, a slab friction factor and the slab length.

Equation 7.2 AASHTO [1] reinforcement detailing equation

$$P_s = \frac{50\,LF}{f_s}\text{(imperial units)} \quad \text{or} \quad \frac{1.15\,LF}{f_s}\text{(metric units)} \tag{7.2}$$

P_s = percentage steel reinforcement
L = slab length, the distance between free edges (feet or m)
F = friction factor at the base of the slab, 1.8 for lime, cement or bitumen stabilised material, 1.5 for unbound gravel or crushed stone and 0.9 for natural subgrade.
f_s = ultimate fracture stress of reinforcement used in the slab (psi or MPa)

If the relationship is applied to a typical 250 mm thick slab, using Grade 250 steel, over either of the standard support platforms, the equation suggests that 200 mm²/m of steel is needed for a 10 m long concrete bay. In this particular case A393 mesh reinforcement would be suitable as the main slab reinforcement. This will therefore easily accommodate any minimum cracking requirement.

Reinforcement design for CRC pavements

As for JRC, the structural contribution from the CRC steel reinforcement is ignored and the AASHTO design method uses the URC calculation (Equation 7.1) to derive a concrete slab thickness. The AASHTO guide then offers the following method for designing the longitudinal reinforcement requirements:

1 The concrete working stress can be estimated using a multi-layered linear elastic method.
2 Equations 7.1 and 7.2 are used to produce the following characteristics within the pavement slab:

    a   a crack spacing of between 1 and 2.5 m;
    b   a maximum crack width of 1 mm;
    c   a maximum reinforcement stress of 75% of ultimate tensile strength.

Transverse reinforcement is designed using the same method as for JRC transverse reinforcement, described in the Section 'Reinforcement design for JRC pavements'.

The following material quantities are recommended to represent standard UK pavement design options:

- Concrete tensile strength at 28 days derived from:

$$f_t = R_T S'_c$$

f_t = indirect tensile strength, mean 28-day value used in design
S'_c = flexural strength (i.e. modulus of rupture), mean 28-day value
R_T = aggregate factor taken as: gravel = 5/8, crushed rock 2/3 [for a standard UK 40 MPa concrete mix, S_c = 5.9 MPa (863 psi) and f_t = 3.69 MPa (570 psi) for gravel or 3.93 MPa (607 psi) for crushed rock]

- Concrete shrinkage at 28-days is derived from the indirect tensile strength.

Indirect tensile strength (psi)	Shrinkage strain
300	0.0008
400	0.0006
500	0.00045
600	0.0003
700	0.0002

Note
A value of 0.0003 can be taken for a UK 40 MPa mix.

- Thermal expansion coefficient ratio (steel/concrete – $\alpha_s\alpha_c$) = 1.32 for limestone aggregate, 0.8 for gravel aggregate.
- Design temperature drop DT_D: 13°C. This value is the difference between the assumed temperature when the concrete is placed and the estimated winter mean temperature.

Thus, for a typical 200 mm thick concrete pavement using Grade 460, 16 mm diameter deformed bar reinforcement. Working stress 1.37 MPa (200 psi) for an 11 tonne (24.3 kip) axle. The standard gives 0.5% reinforcement for a crack spacing of 2.5 m (8.2 ft) and 0.3% reinforcement for a 1 mm (0.04 in) crack width, and also 0.5% reinforcement for a maximum reinforcement stress of 345MPa (50.5 ksi).

It may therefore be concluded that the UK's standard 0.6% reinforcement is sufficient.

7.3 AASHTO interim guide 1998 design method [2]

The current US thinking on pavement design is that the 1992 standard design method may not truly reflect requirements when extreme climatic conditions are encountered. A more complex calculation method is therefore available if required. The method considers a number of different pavement distress conditions individually and then presents calculated pavement thickness values for set design cases. A specific calculation is presented for pavement warping, associated with strong sunlight. The following thicknesses, as described in Table 7.2 apply to two standard design cases.

The AASHTO interim guide [2] further suggests that the design for a 4.6 m long URC slab is appropriate for both JRC and CRC systems.

7.4 TRL research report 87 [3]

Transport research laboratory report RR87 has been mentioned more than once already. It describes the current UK approach to pavement design and the background to the URC and JRC designs in HD 26/01 [4], the UK national design standard. As noted, the design approach is purely empirical, based on the evidence gained from monitoring the performance of a significant number of UK highways. As far as the user is concerned, the

Table 7.2 Summary of AASHTO 1998 [2] interim guidance designs

Design traffic loading (msa)	Concrete slab thickness (m)	
	Support platform A	Support platform B
	$k = 112$ MPa/m; 250 mm granular sub-base 3% CBR subgrade	$k = 662$ MPa/m; 200 mm CBM sub-base 3% CBR subgrade
5	0.223	0.160
10	0.248	0.193
20	0.273	0.228
50	0.305	0.275
100	0.330	0.310

Notes
Other quantities assumed
S'_c = 28-day modulus of rupture (i.e. flexural strength), taken as 5.9 MPa (863 psi) for 40 MPa concrete
$S_o = 0.39$
$k_{subgrade} = 27$ MPa/m (100 pci)
E (granular sub-base option) = 172 MPa
E (CBM sub-base option) = 7,000 MPa
Joint spacing = 5 m
Pavement temperature adjustment = 2.8°C (5°F)
P_t = Terminal serviceability condition = 2.5.

method is simple, consisting of the application of a single equation each for URC and JRC options. Concrete is characterised by a 28-day cube compressive strength, traffic is in millions of standard (80 kN) axles, and the only more difficult quantity is the equivalent foundation modulus (refer Chapter 6). However, the document contains plenty of guidance as to the values appropriate to different foundation designs.

Design for URC pavements

The expression given in RR87 can be adjusted to the following form for design:

Equation 7.3 RR87 [3] design equation for URC pavement

$$H = 0.85 \times 1.15 \left[e^{((40.78 - 3.466\mathrm{Ln}\ S - 0.4836\mathrm{Ln}\ M - 0.08718\mathrm{Ln}\ F + \mathrm{Ln}\ L)/5.094)} \right]$$

$$(7.3)$$

Ln = Naperian (or natural) logarithm
L = pavement life in msa
H = pavement thickness in mm
S = 28-day mean cube compressive strength in MPa
M = equivalent foundation modulus beneath the concrete slab in MPa (see Chapter 6)
F = % failed bays (30% taken as terminal serviceability condition)
0.85 = adjustment to allow for tied shoulders
1.15 = adjustment to produce a 90% level of design confidence

The relationship assumes:

- pavement edges are tied (with at least 1 m concrete edge strip) and untrafficked; otherwise 0.85 is replaced by 1.0;
- a 90% confidence level is assumed; at 50%, 1.15 is replaced by 1.0;
- all joints contain dowels or tie bars;
- the water table is maintained at least 600 mm below the underside of the sub-base.

The following design thickness values, given in Table 7.3, may be calculated from standard UK 40 MPa air-entrained concrete.

The UK standard borrows an adjustment from the US standard [2], reducing the pavement thickness by 15% to allow for the additional support provided by a tied shoulder. Figure 7.1 illustrates the difference between designs using AASHTO and RR87, as well as that from the Cement and Concrete Association document TR550 (see the Section 'TR550 design of floors on ground'), comparing the tied shoulder conditions.

Table 7.3 Summary of RR87 [3] calculated thicknesses

Traffic (msa)	Concrete slab thickness (m)		
	Support platform A	Support platform B	HD 26/01 [4]
	M = 78 MPa; 250 mm granular sub-base; 3% CBR subgrade	M = 489 MPa; 200 mm CBM sub-base; 3% CBR subgrade	M = 270 MPa Std. foundation 150 mm CBM; 5% CBR subgrade
0.5	0.114	0.096	0.150
1	0.131	0.110	0.150
5	0.180	0.151	0.150
10	0.206	0.173	0.160
50	0.282	0.237	0.230
100	0.323	0.272	0.260

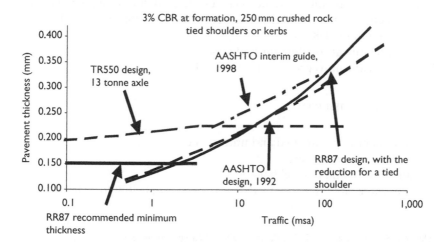

Figure 7.1 URC systems designed over a granular sub-base.

It is noted that common UK practice is to target a 50 MPa value for the concrete strength representing a characteristic 28-day cube strength. The average 28-day compressive cube strength of the PQC is typically around 60 MPa.

Design for JRC pavements

The reinforced concrete predictive relationship presented in RR87 [3] has been adopted and used to generate designs for the UK national standard [4]. The relationship is used in precisely the same manner as for URC except that reinforcement quantity is included as a variable but the percentage number of failed bays is not. The equation in RR87 may be re-expressed as follows:

Equation 7.4 RR87 [3] design equation for JRC pavement

$$H = 0.85 \times 1.15 \left[e^{((45.15 - 3.171\text{Ln } S - 0.3255\text{Ln } M - 1.418\text{Ln } R + \text{Ln } L)/4.786)} \right]$$

(7.4)

Ln = Naperian (or natural) logarithm
 L = pavement life in msa
 H = pavement thickness in mm
 S = 28-day mean compressive cube strength in MPa
 M = equivalent foundation modulus beneath the concrete slab in MPa (see Chapter 6)
 R = is the amount of reinforcement in mm^2/m measured as cross-sectional area of steel per metre width of slab; high tensile steel is used in the UK
0.85 = adjustment to allow for tied shoulders
1.15 = adjustment to produce a 90% level of design confidence

Equation 7.4 assumes:

- pavement Edges are tied and untrafficked;
- a 90% confidence level;
- all joints contain dowel bars;
- the water table is maintained at least 600 mm below formation level.

Table 7.4 Summary of RR87 JRC designs with A393 reinforcement

Traffic (msa)	Concrete slab thickness (m)		
	Support platform A	Support platform B	HD 26/01 [3]
	$M = 78$ MPa; 250 mm granular sub-base; 3% CBR subgrade	$M = 489$ MPa; 200 mm CBM sub-base; 3% CBR subgrade	500 mm^2/m steel $M = 270$ MPa; 150 mm CBM; 5% CBR subgrade
0.5	0.100	0.089	0.150
1	0.116	0.102	0.150
5	0.162	0.143	0.150
10	0.188	0.166	0.150
50	0.263	0.232	0.210
100	0.304	0.268	0.230

Table 7.4 summarises the following design thickness values may be calculated for standard UK 40 MPa air-entrained concrete.

Figures 7.2 and 7.3 illustrate the effect of reinforcement (A393 mesh) according to RR87. The saving in comparison to URC is typically 10–20 mm over a granular foundation, decreasing to less than 10 mm for a CBM sub-base. Lightly reinforced pavement sections therefore appear to perform in an essentially similar manner to an unreinforced pavement. The AASHTO assertion that the reinforcement simply holds together the cracked pavement section appears to be broadly supported.

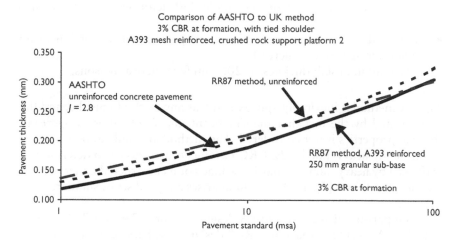

Figure 7.2 JRC systems over a granular sub-base.

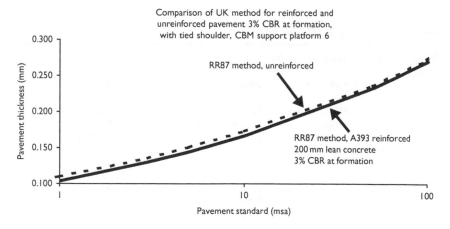

Figure 7.3 JRC systems over a CBM sub-base.

Designs for CRC pavements

The background to the standard UK designs is described in papers [5,6]. After evaluation of the empirical evidence, it was decided to adopt exactly the same equations as presented in RR87 for JRC pavements, with a steel quantity appropriate to CRC. The standard itself [4] simplifies the design process considerably by fixing the following quantities.

- The concrete is a standard 40 MPa mix.
- The longitudinal reinforcement is 0.6% of the concrete sectional area, using 16 mm Grade 460 deformed bar reinforcement (transverse reinforcement is 12 mm diameter Grade 460 deformed bar at 600 mm centres).
- Longitudinal joints are at a maximum spacing of 6 m (or 7.6 m for limestone aggregate concrete).
- The minimum slab thickness is 200 mm for practical reasons.

The UK method is inevitably approximate, as any method must be, and this is illustrated by the fact that the significant impact of different aggregate thermal properties is not considered. The method also fails to take account of unusually heavy axle loads. CRC pavements are known to deteriorate if trafficked by heavy, non-standard axle loads or if constructed under unusual weather conditions [7]. Table 7.5 summarises a typical set of standard CRC designs.

A comparison of the UK and US designs (Figures 7.4 and 7.5) suggests that the AASHTO approach to CRCP design is probably conservative. The US methods fail to take account of the significant contribution presented by the reinforcement. On the other hand, the UK approach, based around a

Table 7.5 Summary of RR87 CRC designs

Traffic (msa)	Concrete slab thickness (m)		
	Support platform A	Support platform B	HD 26/01 [4]
	M = 78 MPa; 250 mm granular sub-base; 3% CBR subgrade	M = 489 MPa; 200 mm CBM sub-base; 3% CBR subgrade	M = 270 MPa; 150 mm CBM sub-base; 5% CBR subgrade
5	0.131	0.122	0.200
10	0.147	0.137	0.200
50	0.191	0.177	0.200
100	0.214	0.198	0.200
500	0.278	0.256	0.270
1,000	0.310	0.288	0.300

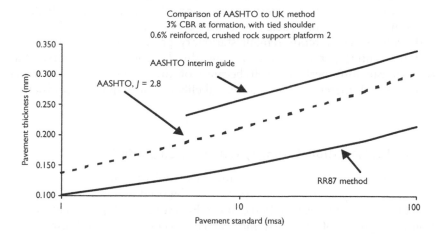

Figure 7.4 CRC systems over a granular sub-base.

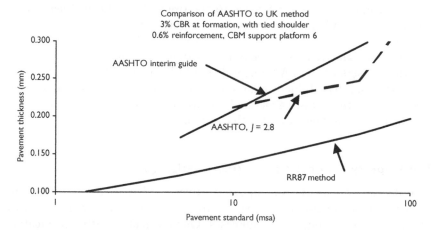

Figure 7.5 CRC systems over a CBM sub-base.

fixed assumption of aggregate thermal properties, can, in some exceptional circumstances, lead to premature pavement failures [7]. It is therefore suggested that both design approaches include some issues that may benefit from further consideration.

7.5 Airfield pavement designs

Concrete is ideally suited to airfield pavements because of the high tyre pressures used on aircraft wheels (typically 1.0–1.5 MPa for large commercial aircraft) and the likelihood of oil spillages occurring. In practice, this means that apron areas, where aircraft stand for loading and refuelling, are almost always constructed in concrete. The same is true of runway end areas, where aircraft remain stationary prior to take-off. Taxiways and the central parts of runways can also be concrete of course, but Asphalt can also perform well because of the greater speed of travel involved and therefore the reduced likelihood of Asphalt deformation occurring.

The Federal Aviation Authority (FAA) design guide [8]

The FAA design documents are widely used all over the world and are particularly easy to apply since they were digitised as the software package LEDFAA. LEDFAA is based on a FE estimate of pavement conditions. The basis of the FAA method is a combination of Westergaard analysis and empirical evidence, the result produces a highly practical and widely trusted method. For practical use the LEDFAA package is recommended in preference to the original graphical design guide charts in the FAA design documents. Figure 7.6 illustrates the design process.

The FAA standards also specify joint spacing is a function of pavement thickness, Users of the FAA method should be aware that the experience upon which it is based chiefly relates to granular or weakly stabilised foundations and that thicknesses generated for pavements on strong cement bound foundations should be treated with caution.

The Property Services Agency (PSA) design guide [9]

This document, shortly to be updated, is very widely used in the UK. The designs, which are empirically based, are expressed in a user-friendly way and cover a wide range of design cases. One particularly attractive feature for airfield engineers is that aircraft load is related directly to the so-called 'aircraft classification number' (ACN), a number used by the industry to describe the damaging effect of each type of aircraft. This means that a PSA design for an aircraft of a certain ACN allows the designer to state that the resulting pavement has an equivalent 'pavement classification number' (PCN), which is the number required by the industry in classifying allowable pavement utilisation.

The flexibility of the PSA guide means that it is well suited to evaluation and rehabilitation design. It has to be borne in mind, however, that in using it in this way, a designer is almost certainly stepping outside the range of evidence upon which the guide is based; it should not, therefore, be used uncritically for such purposes.

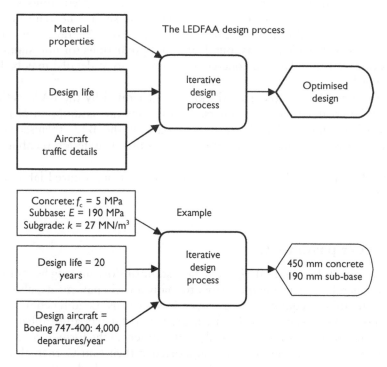

Figure 7.6 FAA designs for Boeing 747-400 loading.

Further points noted with respect to the PSA guide are as follows:

* The effect of cement bound (dry lean concrete) base is taken to be one-third that of the overlying concrete slab, with a maximum contribution to equivalent slab thickness of 50 mm.
* Joints are assumed to be plain undowelled.
* No decrease in pavement thickness is permitted due to concrete reinforcement.
* The concrete strength is the equivalent 28-day flexural strength matched back to a 28-day cored compressive strength.

The British Airports Authority (BAA) guide [10]

The BAA have produced their own, analytically based, design guide. The computation carried out in producing the guide was multi-layer linear elastic, that is, ignoring the effect of the joints and treating the concrete slab as a continuous layer. However, since the relationship between the computed stress to flexural strength ratio and the number of load applications to failure is based on evidence from actual airfields, the predictions should still be reasonably trustworthy. Nevertheless, it should be remembered that any such calibration is really only valid over the range of pavements for which evidence was found, and that predictions for non-standard cases (unusual foundations, concrete strengths or joint details) may be less trustworthy.

In practice, the concrete slab thicknesses generated by the BAA method tend to be greater than most others. This reflects the assumption of plain undowelled joints and a desire on the part of the BAA themselves to avoid in-service maintenance as much as possible. The design reliability is therefore high.

The design method uses the 28-day beam test from a cured laboratory tested sample.

7.6   Port pavements

Port pavements have to be designed to take static or slow-moving heavy loads and are therefore in a similar category to many airfield pavements, although the serviceability requirements are much less onerous. Loading comes from specialist vehicles such as fork-lift trucks, reach stackers and rubber-tyred gantry cranes, and also from stacked goods, largely in the form of containers nowadays. Standard containers are supported at their four Corners through relatively small feet, which means that the pressure applied to the pavement can be very high indeed, particularly where containers are stacked three or four feet high.

The combination of high pressure and slow movement, means that Asphalt is likely to deform, leaving concrete as the most logical pavement design option, although block paving is also possible.

The most commonly used port pavement design manual in the UK is the British Ports Association manual [11,12]. This is based on empirical evidence of performance combined with Westergaard derived calculation, and the basic design is carried out in terms of a standard material, namely a C10 concrete. Other materials are related to C10 concrete by means of an equivalence factor, including options for reinforced concrete, either using conventional steel bar or fibres. The foundation design is based on UK Highways Agency recommendations.

Because, the designs are calibrated using empirical evidence of port pavement performance, they should be reasonably trustworthy for standard cases. However, this is not necessarily true where non-standard loading or non-standard pavement designs are proposed, and in such cases caution is advised. It is always possible to supplement use of the manual with an independent calculation of stress, using either Westergaard or Meyerhof or even multi-layer linear elastic computations.

7.7   Other Westergaard-derived methods and TR550 [13]

All the most commonly used methods of concrete pavement design are based around the Edge and Corner loading Westergaard solutions for the calculation of stress. The estimated in-service load condition is applied and a stress calculated; this is then compared with an allowable working stress based on a standard flexural strength derived for the concrete mix. Such calculations are generally considered to be reasonably reliable but lack the supporting evidence which underlies the more sophisticated highway or airfield design methods. The technique can be particularly useful when exceptionally large or unusual loading conditions have to be designed against.

TR550 design of floors on ground [13]

This UK-published Cement and Concrete Association design method is commonly used in Britain. The following key steps are used with the method.

Design stress

The pavement stresses are calculated using:

- Westergaard internal loading stress   Equation 5.3
- 85% of Kelly's empirical Edge loading stress   Equation 5.7 × 0.85
- 70% of Pickett's empirical Corner loading stress   Equation 5.10 × 0.70

The support offered from adjacent slabs and dowel bars are taken into account by reducing the calculated Edge stress to 85% and the Corner stress to 70% of the calculated values. However, any engineer thinking of using this method is alerted to the fact that the versions of the standard stress calculations given in TR550 [13] do not match the versions given in the original papers.

Subgrade strength

A standard highway based system of estimating subgrade support is used based on the characteristic equilibrium moisture content CBRs as given in the Highways Agency manual [14]. The values are then equated to k_{762} values using Equation 6.6.

Allowable stress based on the concrete mix properties

The calculations are based around characteristic 28-day flexural strength values. The PCA flexural strength to load applications relationship (Equation 4.14) is used to derive the allowable pavement tensile stress.

Table 7.6 TR550 [13] design examples

Number of load applications	Equivalent traffic (msa)	PCA allowed stress (MPa)	Design thickness (m)
25,000,000	174.32	2.25	0.226
4,000,000	27.89	2.25	0.226
500,000	3.49	2.25	0.226
200,000	1.39	2.40	0.217
100,000	0.70	2.51	0.211
57,000	0.40	2.60	0.206
11,000	0.08	2.86	0.195
1,000	0.01	3.25	0.180

Wheel load

This is taken as the heaviest wheel load that is regularly applied to the pavement.

Example calculation

The following example, as described in Table 7.6, calculation is presented as a typical case.

- 13 tonne axle load, that is. 6.5 tonne wheel loads, 7 standard axles per load application;
- mean 28-day flexural strength 6 MPa;
- characteristic 28-day flexural strength 5 MPa;
- $k = 96$ MPa/m, that is, a granular pavement foundation.

In situ concrete industrial hard-standings [15]

A second version of this method can be found within this well presented guide. The book presents the standard TR550 design method but additional load factors are applied depending on the traffic type. The guide is considered to produce more reliable designs when compared with the standard TR550 approach.

7.8   Other Meyerhof-derived methods and TR34 [16]

The limit state approach taken by Meyerhof is well suited to statically loaded floor slabs, which can only fail by total collapse rather than by the propagation of cracks due to fatigue under multiple loading. For this reason, it forms the principal computational tool behind the UK Concrete Society publication TR34, the standard currently used for the design of internal industrial warehouse and floor systems in the UK. The document is specifically aimed at designing concrete slabs for large warehouse projects where high-quality pavement finishes are needed and the subgrade is not affected by a freeze–thaw cycle. The design method is intended for internal warehouse slabs and cannot therefore be directly related to external pavement designs. Pavement thickness design is based on Meyerhof and the Portland Cement Association fatigue model described in TR550.

7.9 Discussion and recommendations

There is every reason to place reasonable trust in all the major published design approaches; after all each has been compiled by professionals with considerable experience of concrete pavement performance. This means that whether Westergaard, Meyerhof or an adaptation of multi-layer linear elastic theory has been used, or the basis is purely empirical, one can be reasonably certain that sensible calibration has been built into the procedure, ensuring that predictions match experience. However, it must be remembered that each method is therefore only as good as the evidence and experience upon which it is based. It makes sense to trust the FAA designs for URC pavements under aircraft loading, for example, but not to start applying them to highway loading or industrial floor applications. With this point in mind, therefore, the following observations are offered on the strengths and weaknesses of the different methods.

Road pavements

As noted already, there is little difference between the AASHTO and RR87 designs for URC pavements on granular sub-bases, which is as one would expect since there is plenty of past experience available. However, the difference increases slightly when cement bound sub-base is introduced, and it may be that the RR87 designs are slightly more trustworthy since this type of pavement construction has been in widespread use in the UK for several decades, informing the TRL in the development of the method. It is also undeniably true that the RR87 method is the simpler. Similarly, it seems reasonable to place a certain trust in the reduced thicknesses advised by RR87 (but not by AASHTO) for JRC and CRC constructions.

However, serious questions only really begin to arise in non-standard cases, situations which lie outside of the experience upon which the design guides are based. Typically, this might involve ultra-heavy loading, unusually strong (or weak) concrete, or non-standard bay dimensions, and in such cases it is not logical to expect a purely experience-based method such as RR87 to give a correct answer. A method such as AASHTO, which is based on a sound theoretical analysis (Westergaard) may be expected to cope better – but not in matters to do with traffic, since the AASHTO manual artificially reduces traffic to standard axles in the same way as RR87. Looking ahead to the planned publication of the AASHTO 'mechanistic' design guide, which will take detailed account of traffic, here it may be possible to place some trust in the output even for non-standard load cases.

In short, for standard major highways, the equations in RR87 are considered both simple and trustworthy, certainly for UK conditions. The concrete strength parameter should be fixed around the characteristic strength of the planned pavement concrete to reliably represent the planned pavement design. The pavement should be kerbed or have untrafficked (tied) shoulders. Joints should be dowelled. For some non-standard cases, AASHTO is likely to be more useful since it has a sound theoretical basis. However, never be afraid to apply theoretical computations from first principles, so long as the results are not simply believed but are 'anchored' back to experience-based designs. In really non-standard cases (e.g. multiple close-spaced wheels) this may be the only way forward.

Airfield pavements

Very similar comments apply to aircraft load designs. Each design manual relates to the experience upon which it is based. This means dowelled joints in the case of the FAA method, undowelled in the BAA method, reflecting the different policies of the two organisations; and the inevitable consequence is that BAA designs are thicker (if all other design parameters remain the same). For major airfield pavement design in the UK, the BAA designs are easy and safe, but they may be less appropriate where different policies are adopted (different concrete strengths or joint spacing, JRC pavements). In the authors' opinion, the PSA method is a very useful combination of sound experience-based designs, flexibility (for instance in dealing with rehabilitation) and adequate simplicity.

However, the word 'non-standard' is much more likely to apply to airfield pavements than to roads and for this reason a well-rounded engineer should certainly be prepared to analyse from first principles where necessary. The combined effect of four or even six wheels should not be ignored and this means that, when stepping outside the experience base of the established design methods – for example in connection with the new Airbus A380 – it is sensible to conduct one's own analysis (Meyerhof is likely to be of most use) so long as it ties in with the more expected designs for standard cases.

Industrial pavements

Here too, methods such as TR550 or TR34, both of which combine analysis (Westergaard and Meyerhof, respectively) with experience, are believed to be proven and therefore appropriate for internal floor slabs. The British

Ports Association manual, which is very largely experience-based, is surely reasonably trustworthy for standard port design cases, including container stacks, fork lifts, gantry cranes, etc. However, it is *always* prudent to use more than one method in difficult cases, including a first principles analysis. For example, the British Ports Association manual gives a series of relatively simple factors depending on concrete strength, whether reinforced or not, including a fibre reinforced option, but the very simplicity of the method makes it quite certain that the designs cannot possibly be optimised in every case, and in some cases designs could even be 'unsafe' rather than 'safe'. The message therefore has to be: never believe the result from any manual unquestioningly; always get a second opinion!

7.10   References

1. American Association of State Highway and Transportation Officials, *AASHTO Guide for Design of Pavement Structures*, American Association of State Highway and Transportation Officials, 1992, ISBN 1-56051-055-2.
2. American Association of State Highway and Transportation Officials, *Supplement to the AASHTO Guide for Design of Pavement Structures*, American Association of State Highway and Transportation Officials, 1998, ISBN 1-56051-078-1.
3. Mayhew, H.C. and Harding, H.M., *Thickness Design of Concrete Roads*, Research Report 87, Transport and Road Research Laboratory, 1987, ISSN 0266 5247.
4. The Highways Agency, *Design Manual for Roads and Bridgeworks*, vol. 7, Pavement Design, HD 26/01 Pavement Design.
5. Gregory, J.M., Continuously reinforced concrete pavements, Paper 8773, Proceedings of the Institution of Civil Engineers, London, May 1984.
6. Garnham, M.A., *The Development of CRCP Design Curves*, Highways and Transportation, London, December 1989.
7. Cudworth, D.M. and Salahi, R., A case study into the effects of reinforcement and aggregate on the performance of continuously reinforced concrete pavements, Fourth European Symposium on Performance of Bituminous and Hydraulic Materials in Pavements, BITMAT 4, Nottingham UK, April 2002.
8. International Civil Aviation Organisation, *Aerodrome Design Manual*, Part 3 Pavements, 2nd edn, Quebec, 1982.
9. Property Services Agency, *A Guide to Airfield Pavement Design and Evaluation*, HMSO, 1989, ISBN 0 86177 127 3.
10. BAA plc, Group Technical Services Aircraft Pavements, *Pavement Design Guide for Heavy Aircraft Loading*, R. Lane, London, 1993.
11. Knapton, J., *The Structural Design of Heavy Duty Pavements for Ports and Other Industries*, 1st edn, British Ports Association, 1986, ISBN 0 900337 22.
12. Knapton, J. and Meletiou, M., *The Structural Design of Heavy Duty Pavements for Ports and Other Industries*, 3rd edn, The British Precast Concrete Federation Ltd and The British Ports Association, 1996, ISBN 0 90037 22 2.

13. Chandler, J.W.E., Design of floors on ground, Technical Report 550, Cement and Concrete Association, June 1982.
14. *Design Manual for Roads and Bridges*, vol. 7, section 2, HD 25/94 Subgrade Assessment, London, 1994.
15. Knapton, J., *In-situ Concrete Industrial Hardstandings*, Thomas Telford, ISBN 0-7277-2827-x, 1999.
16. The Concrete Society, *Technical Report 34*, 2nd edn, 1994, ISBN 0-946691-49-5.

Chapter 8

Composite pavement design

8.1 Introduction

The calculations supporting composite pavement design are in some ways confusing, since three quite different approaches may be found within different mainstream design standards, namely:

- An American system described in AASHTO [1] which uses empirical methods developed directly from flexible pavement design.
- A UK method developed by TRL and described by Nunn [2]. This UK approach is essentially empirical but uses a number of calculations borrowed from both flexible and rigid concrete pavement design.
- Scientifically based methods, notably those presented in South African and French standards. These systems use aspects of both rigid and flexible pavement design.

Each approach is described in more detail within this chapter. The scientific principles embedded within the methods are not always precisely identifiable. The most scientifically rational approach is probably that in the French and South African national standards, but it is difficult to compare the methods directly due to the lack of any common basis. The best advice that can be offered is to suggest that each method should be applied within the confines of the adopting authority's specifications. A set of standard calculations will be undertaken for the AASHTO and UK methods using the following standard materials and construction platforms:

- 100 pen dense bitumen macadam (DBM) Asphalt layers;
- CBM layer of 10 MPa 7-day compressive cube strength material;
- crushed rock sub-base material;
- 3% CBR subgrade.

8.2 The AASHTO design method [1]

The standard AASHTO flexible pavement design method can also be successfully applied to composite pavement design. The core of this design method is Equation 8.1, which relates a quantity known as structural number (S_n), a measure of overall pavement strength, to traffic level, subgrade condition and parameters relating to design reliability and variability, as well as acceptable condition at the end of the pavement's life. Having established the required structural number, the designer then has to decide the most appropriate materials and thicknesses by which to attain that level of pavement strength. This is achieved in a simplistic but practical way by using Equation 8.2.

Equation 8.1 AASHTO [1] basic equation for flexible pavement design

$$Log_{10} W_{18} = Z_R S_0 + 9.36 Log_{10}(S_n + 1) - 0.20$$

$$+ \frac{A}{B} + 2.32 Log_{10} M_R - 8.07 \tag{8.1}$$

$$A = Log_{10} \left(\frac{\Delta PSI}{4.2 - 1.5} \right) \tag{8.1a}$$

$$B = 0.40 + \left(\frac{1094}{[S_n + 1]^{5.19}} \right) \tag{8.1b}$$

W_{18} = traffic loading in standard axles
Z_R = adjustment to give different levels of confidence in design
M_R = subgrade surface modulus in psi
S_0 = the standard deviation of the data used to construct the pavement, taken as 0.35
ΔPSI = serviceability loss (see Chapter 7, the Section 'Concrete thickness design')
S_n = pavement 'structural number', a measure of the intended pavement strength

Equation 8.2 AASHTO [1] equation to estimate layer thickness

$$S_n = S_{n_1} + S_{n_2} + S_{n_3} = a_1 D_1 + a_2 D_2 m_2 + a_3 D_3 m_3 \tag{8.2}$$

a_1, a_2 and a_3 are layer coefficients (see later)
D_1, D_2 and D_3 are layer thicknesses in inches
m_1 and m_2 are moisture coefficients

Basically, the required structural number is made up of contributions from an Asphalt surfacing, a base and a sub-base, and these contributions are a function of thickness, stiffness and, for granular materials, drainage provision. The procedure is first to use Equation 8.1 with M_R as the base modulus, giving a required structural number for the contribution of the Asphalt. The step is then repeated with M_R as the sub-base modulus, yielding a combined Asphalt and base structural number. Since the Asphalt contribution has already been determined, this leaves the base contribution. Finally, with M_R as the subgrade modulus, the total structural number can be calculated, allowing the contribution of the sub-base to be determined.

The following steps relate to an example computation with a cement bound base layer.

1   The serviceability loss ΔPSI is estimated in a similar manner to the rigid pavement calculation in Chapter 7. The initial serviceability P_0 may be assumed as 4.5. The terminal serviceability P_t can be taken as 2.5 for high-quality roads. This would give a ΔPSI value of 2, but from this must be subtracted the serviceability loss due to frost, subgrade heave and drainage deterioration. A well-drained UK pavement over a well-drained subgrade constructed in non frost-sensitive material would probably give a serviceability loss of around 0.35 in 20 years, giving a final ΔPSI value of 1.65.

2   The pavement layers are characterised according to their stiffness moduli, from which layer coefficient values are then derived using correlations provided in the AASHTO manual. The following are typical values suggested for UK materials.

Layer	Material description	Stiffness	Layer coefficients
Asphalt	100 pen DBM	3 GPa (0.4 mpsi)	$a_1 = 0.42$
CBM base	10 MPa at 7 days	5 GPa/500 MPa[a]	
		(0.7 mpsi/70,000 psi)	$a_2 = 0.27$; $m_2 = 0.8$
Sub-base	30% CBR	100 MPa (15,000 psi)	$a_3 = 0.11$; $m_3 = 1.2$
Subgrade	3% CBR	30 MPa (4,500 psi)	—

Note
a The lower value is a long-term design value, accounting for stress concentrations at cracks.

3   The moisture coefficients for each layer are estimated based on tables given in the manual. Values typically range from 0.7 to 1.25. For example,

0.75 = poorly drained, wet for 25% of the year;
1.0 = fairly well-drained, wet from 1% to 5% of the year;
1.25 = excellent, wet for less than 1% of the year.

4   The subgrade strength is then estimated, see Chapter 6; if a 3% CBR is assumed, the roadbed effective modulus becomes:

$$M_R = \text{CBR} \times 1,500 = 4,500 \text{ psi}.$$

5   Each layer's structural number will then be evaluated in turn, using Equation 8.1. The S_n values are calculated with the selected standard deviation, $S_0 = 0.35$, traffic load and foundation modulus, where each S_n value is based on the stiffness of the layer immediately below.

   • Asphalt layer, S_{n_1}, based on CBM stiffness of 70,000 psi (500 MPa) as the M_R value (note: this is the residual long-term value).
   • Combined Asphalt and CBM layer, S_{n_2}, based on sub-base stiffness of 15,000 psi (100 MPa) as the M_R value.
   • Total pavement construction, S_{n_3}, based on subgrade stiffness of 4,500 psi (30 MPa) as the M_R value.

6   The thickness of each layer is then derived using Equation 8.2.
7   Each layer's design thickness is then adjusted to represent a realistic practical pavement construction option. Table 8.1 gives the results of sample calculations.

A standard 50 msa pavement design would therefore have the following construction:

   • DBM surfacing, 100 pen material   120 mm.
   • CBM base material, 10 MPa mean cube strength at 7 days   190 mm.
   • Crushed rock sub-base, 30% CBR material   380 mm.

Table 8.1 AASHTO [1] derived composite pavement thicknesses

Traffic (msa)	Asphalt layer $a_1 = 0.42$		CBM base layer 10 MPa at 7-days $a_2 = 0.27; m_3 = 0.8$		Sub-base layer 30% CBR $a_3 = 0.11; m_3 = 1.2$	
	S_{n_1}	Thickness (mm)	S_{n_2}	Thickness (mm)	S_{n_3}	Thickness (mm)
1	1.0	60	1.9	110	3.0	210
5	1.3	80	2.5	140	3.9	270
10	1.5	90	2.8	150	4.4	310
20	1.7	100	3.1	160	4.9	350
50	2.0	120	3.6	190	5.6	380

8.3 UK standard designs

The UK design standard can present a wide range of possible options depending on the base material selected (including the use of alternative binders such as fly ash and slag) and the foundation class. Details are given in HD26/05 of Volume 7 of the Highways Agency's Design Manual for Roads and Bridges [3]. The following paragraphs outline the key elements relating to each pavement layer.

Bituminous surfacing

The thickness of bituminous surfacing is dictated by the need to provide a protective layer over a pre-cracked CBM. For heavily trafficked highways, the minimum value has been chosen to provide protection against reflective cracking of the bituminous layer over the cracked CBM material.

CBM base

The CBM base layer thickness depends on the grade of material used. This is based on the 7-day cube strength, which is linked to the 28-day flexural strength. However, rock aggregate CBM materials are known to have higher flexural strengths than washed river gravels (for the same compressive strength); thus a 'CBM3R' design is thinner than 'CBM3G'.

Foundation

The standard allows four different classes of foundation, linked to the equivalent foundation modulus expected in the long term. As a rough guide:

- Class 4: strong CBM sub-base;
- Class 3: weaker CBM (or slag or fly ash bound) sub-base;
- Class 2: crushed rock sub-base;
- Class 1: sand/gravel sub-base.

HD 26/01 contained a summary of the different pavement design thicknesses for each of the different materials.

A 50 msa design utilising 10 MPa CBM over crushed rock sub-base would be:

- DBM surfacing, 100 pen material   175 mm.
- CBM base material, 10 MPa mean cube strength at 7 days   190 mm.
- Crushed rock sub-base, 30% CBR material   300 mm.

This is clearly significantly different from the AASHTO design, with 55 mm more Asphalt, the same CBM thickness and 80 mm less sub-base. The principal difference in philosophy responsible for this is the UK's perceived need to protect the surfacing against reflective cracking.

8.4   UK Technical Paper PE/TP/169/92 [2]

The UK composite pavement design method was the basis for the current Highways Agency standard [3] and used a combination of plastic analysis and an adjusted Westergaard calculation to derive standard pavement designs. The criterion for acceptance was that the ratio of flexural strength to the combined CBM thermal and traffic-derived stress must not fall below 1.5. Nunn's unpublished report [2] describing the background to the original design method indicates that two alternative approaches were considered. In pavements carrying up to 20 msa the base material is allowed to deteriorate under thermal and traffic loading. Over 20 msa and up to 80 msa the base is designed to be stable and not to deteriorate under traffic loading.

The report suggests that deterioration is a function of:

- the thickness of bituminous overlay reducing thermal stress;
- traffic stress in the CBM layer;
- the 28-day flexural strength of the CBM material.

The report estimates the following variables and examines their influences on pavement design:

- thickness of bituminous overlay;
- thermal properties of CBM;
- CBM strength and thickness;
- distance between cracks in the CBM;
- foundation stiffness.

The calculation method is summarised by the following steps:

1   The allowable ratio of 28-day mean flexural strength to combined traffic and thermally induced stress is as follows.

Equation 8.3 Design stress in CBM materials

$$f_{t,fl} \leqslant 1.5 \, (\sigma_{tr} + \sigma_{therm}) \tag{8.3}$$

$f_{t,fl}$ = the 28-day mean flexural strength
σ_{tr} = the traffic-induced stress from a 40 kN wheel load
σ_{therm} = the thermally induced warping stress at the underside of the CBM layer

2 Traffic stress calculation: a multi-layer elastic model is used to calculate the stress at the underside of the CBM layer induced by a standard 40 kN, 0.151 m radius patch load.

Stiffness assumptions:

- Asphalt – top 80 mm = 0.9 GPa; next 40 mm = 1.9 GPa; remainder = 3.1 GPa;
- CBM – stiffness modulus = $(\text{Log}_{10}f_{\text{fl},28\text{-day}} + a)/b$ (where 28-day strength = 1.1 × 7-day strength; $f_{\text{fl}} = c \times f_{\text{c,cube}}$; for gravel aggregate, $a = 0.773, b = 0.0301, c = 0.11$; for crushed rock aggregate, $a = 0.636, b = 0.0295, c = 0.16$). For example: "CBM3G" = $(\text{Log}_{10}f_{\text{fl},28\text{-day}} + 0.86)/0.0301 = 29$ GPa.

Example stress calculations – after [2]:

Bituminous layer thickness (mm)	CBM3G layer thickness (mm)	Sub-base layer thickness (mm)	Subgrade strength (CBR) (%)	CBM layer stress σ_{tr} (MPa)
190	150	250	3	1.09
190	200	250	3	0.82
190	250	250	3	0.63
190	300	250	3	0.49

3 Thermal warping stress: This is calculated using a method developed by Thomlinson [4], as follows:

Equation 8.4 Thermal warping stress calculation after Thomlinson [4]

$$\sigma_{\text{therm}} = \frac{E\alpha\theta}{(1-v)}A_1K_1 = 1.64 \times 10^6 A_1 K_1 \tag{8.4}$$

σ_{therm} = thermally induced stress
E = CBM stiffness modulus, taken as 29 GPa for CBM3G
α = coefficient of CBM expansion, taken as $12 \times 10^{-6}\,^\circ\text{C}^{-1}$ for all standard calculations
θ = amplitude of diurnal temperature variation, taken as 4°C for 200 mm bituminous layer or 5.3°C for 150 mm bituminous layer when subjected to a 10°C temperature gradient
v = Poisson's Ratio, taken as 0.15
A_1, K_1: see below; A_1 reflects the non-linear nature of the temperature gradient through the CBM slab

Equation 8.5 A_1 adjustment after Thomlinson [4]

Derivation of A_1

$$a_h = \frac{H}{d}\left(\frac{\pi}{T}\right)^{0.5} = H \times 0.0064 \tag{8.5}$$

a_h = thickness coefficient, related to a standard set of factors in Nunn's work

H = composite slab thickness (mm), taken as CBM thickness + 0.1 bituminous thickness

d = diffusivity factor taken as $(0.9)^{0.5} = 0.94$

T = the periodic time cycle = 86,400 s

A_1 depends on a_h as follows [4]:

Thickness coefficient, a_h	Coefficient of amplitude of the stress cycle, A_1
0.5	0.22
1.0	0.37
1.5	0.38
2.0	0.36

Example values of A_1 used in design – after [2]:

Bituminous layer thickness (mm)	CBM3G layer thickness (mm)	Composite thickness H (mm)	a_h	A_1
190	150	169	1.08	0.38
190	200	219	1.40	0.38
190	250	269	1.72	0.37
190	300	319	2.04	0.36

Equation 8.6 K_1 thermal warping stress adjustment after Bradbury [7]

Derivation of K_1

$$K_1 = \left(\frac{C_1 + \nu C_2}{1 + \nu}\right)^{0.5} \tag{8.6}$$

C_1, C_2 = constants derived from the ratio of crack spacing to radius of relative stiffness of the composite CBM and bituminous system

Ratio: CBM crack spacing/radius of relative stiffness	C_1, C_2
3	0.2
4	0.45
5	0.75
6	0.94
7	1.05

Example values of K_1 used in desing – after [2]:

Bituminous layer thickness (mm)	CBM3G layer thickness (mm)	Composite thickness H (mm)	Radius of relative stiffness ℓ (mm)	B/ℓ (mm)	C_1, C_2	K_1
190	150	169	596	5.0	0.72	0.72
190	200	219	724	4.1	0.49	0.49
190	250	269	844	3.6	0.38	0.38
190	300	319	958	3.1	0.20	0.20

Note
Assume $k_{foundation}$ = 96 Mpa/m; CBM crack spacing, $B = 3$ m.

Example thermal stress calculation [2]:

Bituminous layer thickness (mm)	CBM3G layer thickness (mm)	Composite thickness H (mm)	K_1	A_1	σ_{therm} (MPa)
190	150	169	0.72	0.38	0.45
190	200	219	0.49	0.38	0.31
190	250	269	0.38	0.37	0.23
190	300	319	0.20	0.36	0.12

4   Combined stress calculation: the two stresses are added and compared with the flexural strength of the material.

Bituminous layer thickness (mm)	CBM3G layer thickness (mm)	σ_{therm} (MPa)	σ_{tr} (MPa)	$1.5 \times (\sigma_{therm} + \sigma_{tr})$ (MPa)	f_{fl} (MPa)	Is the pavement acceptable?
190	150	0.45	109	2.31	1.20	No
190	200	0.31	0.82	1.69	1.20	No
190	250	0.23	0.63	1.29	1.20	No
190	300	0.12	0.49	0.49	1.20	Yes

Actually, the standard design would include a thicker sub-base layer, 300 mm rather than 250 mm, and this means that the 190 bituminous/250CBM solution becomes acceptable.

This calculation directly parallels that presented in Chapter 5 for concrete surface slabs and, although the sheer number of steps used makes it somewhat confusing, this method forms the basis of all UK Highways Agency flexible composite pavement design.

8.5   The South African method [5]

A substantial body of advice on CBM pavement design exists in the relevant South African standard [5]. The method is the product of many years of research based around tests undertaken using an accelerated pavement testing (APT) machine, which is used to traffic a real pavement to failure. The method has a number of individual elements that are not used in other design methods.

1   Four different categories are applied to the different types of road found in South Africa. The different levels of reliability required are as follows.

Road category	Description	Assumed level of reliability in design (%)
A	Interurban freeway	95
B	Interurban collector and major rural roads	90
C	Rural roads	80
D	Lightly trafficked rural roads	50

2   Cement bound pavements may be designed to three different progressive failure conditions:

a   Phase 1, fatigue cracking in CBM, no crushing is assumed to occur.
b   Phase 2, CBM crushing is initiated.
c   Phase 3, the CBM suffers advanced crushing failure.

(The South African standard allows the use of large quantities of low-strength CBM materials that are susceptible to crushing.)

3   Three complete sets of calculations are then presented to give the estimated life of each layer during each phase of design. A chosen pavement design can then be adjusted to reflect the required service conditions and pavement life.

4   The method allows the use of a 'shift factor' to adjust the chosen fatigue model to reflect the over-design produced by using a thick continuous CBM layer, but regrettably the method fails to suggest how the factor is applied in a pavement design. The shift factor varies from 1.0 at CBM thicknesses below 100 mm to 8.0 at thicknesses over 400 mm.

The calculation uses a standard 40 kN dual wheel load with the tyres spread 350 mm apart with a 0.52 MPa contact pressure. Stresses and strains are calculated using a standard multi-layer linear elastic computation. The Phase 1 calculation is for the maximum strain at the base of the CBM layer. This strain is related to the number of load applications in Phase 1 as follows:

Equation 8.7 South African [5] CBM fatigue life acceptance criteria for Category A roads

$$N_{eff} = 10^{6.721(1-(\varsigma/7.49\varsigma_b))} \tag{8.7}$$

N_{eff} = estimated pavement design life for Phase 1
ς = maximum calculated strain in the pavement
ς_b = default material strain value

The Phase 2 calculation involves the maximum vertical compressive stress at the top of the CBM layers, which is related to the number of load applications within Phase 2 as follows:

Equation 8.8 South African [5] CBM crushing initiation life acceptance criteria for Category A roads

$$N_{ci} = 10^{7.386(1-(\sigma_v/1.09UCS))} \tag{8.8}$$

N_{ci} = estimated design life for Phase 2
σ_v = maximum calculated stress at the top of the CBM layer
UCS = default unconfined compressive strength for the test material

The Phase 3 calculation also makes use of the maximum compressive stress in the CBM. The relevant equation is:

Equation 8.9 South African [5] CBM advanced crushing life acceptance criteria for Category A roads

$$N_{ca} = 10^{8.064(1-(\sigma_v/1.19UCS))} \tag{8.9}$$

N_{ca} = estimated design life for Phase 3
σ_v = maximum calculated stress
UCS = default unconfined compressive strength for the considered material

Further calculations are then undertaken to check the bituminous material fatigue life and subgrade rutting. The South African method provides substantial information on the behaviour and design of CBM materials; it can be a useful tool to designers considering how to analyse a non-standard composite or CBM sub-base pavement.

8.6 The French method [6]

A second high-quality method for design can be found in the French national standard [6]. The method uses two linked sets of calculations. A rational concrete design method is used to determine CBM layer thickness; a second separate design calculation is used to determine bituminous layer thickness. The design method has the following important elements:

1   All of the pavement calculations are undertaken using a 13 tonne axle load, applied using two sets of two wheels. The wheels are set 0.375 m apart with a patch load radius of 0.125 m and a contact pressure of 0.662 MPa.
2   The CBM thickness is designed by computing the CBM stress at the underside of the layer, immediately below the wheel load. The calculations are undertaken using uncracked CBM stiffness values based on the expected in situ stiffness of the layer, nominally 360 days after construction. They assume a full bond between the CBM and bituminous layer. A concrete fatigue model is used to relate stress : strength ratio to pavement life via a series of adjustment factors, as follows:

Equation   8.10 French   [6]   maximum   permitted   tensile   stress
requirement

$$\sigma_{t,ad} = \sigma_t(NE)k_r k_d k_c k_s \tag{8.10}$$

$\sigma_{t,ad}$ = the adjusted strength of the CBM layer
$\sigma_t(NE)$ = bending failure stress (flexural strength) on a sample at 360 days
k_r = adjustment for reliability, the example in the text [6] uses a value of 0.744
k_d = adjustment to reflect joints and cracks in the CBM layer, taken as 1.25
k_c = calibration adjustment to the calculation, taken as 1.4 or 1.5 for different CBM materials
k_s = adjustment for the reduction in bearing capacity due to laying material over weak subgrade, taken as 0.8 for subgrades less than 50 MPa and 1 for material over 120 MPa

3   The thickness of bituminous surfacing is determined using a stress–strain calculation including the effective, cracked CBM layer stiffness value. The effective CBM stiffness value is taken to be the uncracked CBM stiffness value divided by 5. No bond is assumed between the CBM and bituminous layers.

The method has many advantages when compared with other standards except that the calibration and fatigue acceptance criteria are all rolled up into the fatigue acceptance model, which is complex and difficult to understand. In general it is a good scientific method but the system uses a large number of factors to produce an effective design. The application of these adjustment factors could introduce errors into the interpretation and application of the design method especially if the method is used outside of its knowledge base, France, or for different materials, climates, etc.

8.7   Reflective cracking

The issue of reflective cracking has been referred to already, notably as a limiting factor on the bituminous layer thickness in the UK design approach. It is not the purpose of this document to delve into the many and various ways in which reflective cracking can be (a) analysed, (b) predicted and (c) inhibited. However, it must be emphasized that the practice of pre-cracking, usually at intervals of around 3 m, is an essential element in the fight against reflective cracking. This procedure was developed in France and has now been adopted as standard in the UK and several other countries also. Chapter 9 explains the background and the available techniques.

8.8   Discussion

Each of the different national approaches to composite pavement design offers an entirely different methodology. Very little common ground appears to exist between the different design methods and standards. It is recommended that exceptional care is exercised in applying each of the different methods to real pavement designs, since the interpretation of each method is open to differing opinions.

8.9   References

1. American Association of State Highway and Transportation Officials, *AASHTO Guide for Design of Pavement Structures*, American Association of State Highway and Transportation Officials, 1992, ISBN 1-56051-055-2.
2. Nunn, M.E., Factors affecting the long life design of flexible composite roads, Technical Paper PE/TP/169/92, TRL, 1992, on behalf of Britpave and the Highways Agency, unpublished, London.
3. The Highways Agency, *Design Manual for Roads and Bridgeworks*, vol. 7, Pavement Design, HD 26/05, London.
4. Thomlinson, J., Temperature variations and consequent stresses produced by daily and seasonal temperature cycles in concrete slabs, *Concrete and Construction Engineering*, Issue Number 35.
5. Strauss, J.P. and Jordaan, G.J., Rehabilitation design of flexible pavements in South Africa, Department of Transport RR 93/296, South Africa.
6. *French Design Manual for Pavement Structures*, Laboratoire Central des Pontes et Chaussées, Service d'Etudes des Routes et Autoroutes, May 1997, D 9511TA 200 FRF, Paris.
7. Bradbury, R.D., *Reinforced Concrete Pavements*, Wire Reinforcement Institute, 1938, Washington DC.

Chapter 9

Joints

9.1 Introduction

It has already been noted that correctly designed and constructed joints are essential to the efficient operation of concrete pavements. Joints may be designed in a number of different ways and can be either dowelled, undowelled with crack inducers or formed. The detailing of a jointing system is specific to the pavement type and must be incorporated into the pavement design method. Each form of construction is examined separately within this chapter.

9.2 Joint efficiency

Joint efficiency can be measured using an end-product test where the ability of the pavement to transmit load across the jointed gap is recorded. The level of efficiency is described using Figure 9.1 and Equation 9.1. Joint efficiency is reported in terms of a percentage of deflection transmitted across the gap. The test is usually undertaken using a Falling Weight Deflectometer (FWD) but a static plate-loading test may also be used. The minimum level of joint efficiency is different in each design method and can be linked back to the stress/strain calculation technique. The UK [1] minimum level of acceptance for highway pavements is 75% but the standard suggests that 90% or more should be achieved. The Portland Cement Association [2] quotes roller compacted concrete crack efficiencies of between 40% and 60% for saw-cut joints and 60–90% for naturally cracked joints. Recent US papers (e.g. [3]) record a very high joint efficiency of over 90% for CRC. UK airfield pavement design uses a different measure of acceptance; the PSA [4] method uses a variable percentage load transfer, giving values from 5% to 33%. In practice of course, the actual efficiency varies significantly from joint to joint and is also enormously affected by slab temperature, since joint width changes as each concrete panel expands and contracts. The UK Highways Agency advise that joint efficiency should not be measured at a temperature greater than 15°C in order to avoid deceptively high results.

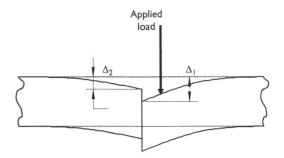

Figure 9.1 Joint efficiency.

Equation 9.1 Joint efficiency

$$\text{Joint efficiency} = \frac{\Delta_2}{\Delta_1} \times 100\% \tag{9.1}$$

9.3 Load transfer devices

Load transfer can be maintained using dowels and tie bars, aggregate interlock or keyways; each method is described separately.

Aggregate interlock

Aggregate interlock is the simplest method of achieving load transfer. It is generally accepted and stated in a paper by Hanekom *et al.* [5], that the width of crack must be maintained below 1 mm to achieve a reasonable level of load transfer across an aggregate interlock joint. This method of load transfer is typically obtained using a crack-induced joint where the joint is formed as part of the construction process. Two different techniques may be used; sawing the concrete slab (within 24 h of construction) or inserting a crack inducer into the wet concrete. Whichever method is used, the sawn or formed joint depth must extend to between 1/4 and 1/3 of the pavement depth to ensure the formation of a clean crack.

Timber triangular crack inducers can be used to assist in the formation of a crack in a concrete pavement. The triangular timber former is placed on the sub-base immediately below the planned joint. However, recent UK experience suggests that this technique should be avoided; the method is unreliable and can lead to the formation of cracks away from the surface joint. Figure 9.2 illustrates the problem.

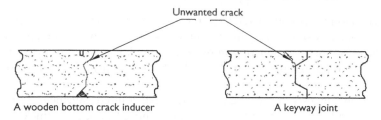

Figure 9.2 Possible joint construction problems.

Keyways

Keyways and sinusoidal joints are forms of construction which are occasionally used in association with mass concrete (URC) pavements instead of relying on aggregate interlock for load transfer. However, practical experience of this type of construction suggests that the joints can become a major maintenance problem. Many reported cases exist which suggest that the upper and lower sections of the keyway can be susceptible to fracture, and similar problems have also been reported with sinusoidal joints. Figure 9.2 illustrates the problems. Neither keyways nor sinusoidal joints are recommended.

Dowels and tie bars

Dowel bars are the commonest load transfer devices; dowels allow horizontal movement to occur between adjacent pavement panels over a gap greater than 1 mm. Dowel bars can be designed by theoretical methods but most pavement design guides use a standardised bar size, steel grade and spacing. A Cement and Concrete Association report [6] summarises the technical background to dowel bar design. The most important issue to consider is to ensure that the dowel bars are inserted into the pavement in a manner that allows movement to occur; one half of an embedded dowel must be de-bonded. Dowel bar lock-up is a common cause of pavement distress. Special de-bonding agents or sleeves are used to ensure free joint movement. Cement and Concrete Association Technical Report 403 [7] describes the background to the identification of suitable de-bonding agents. Bituminous paint is not recommended as a de-bonding agent.

Tie bars are used in longitudinal joints to hold together adjacent strips of concrete, allowing a degree of flexure at the joint but no opening or closing. Tie bars are smaller in diameter than dowels and are embedded and bonded to each side of the joint. The central section of a tie bar, which crosses the joint itself, will require protection from corrosion.

Dowels compared with aggregate interlock alone

Dowelled joints are considered to be more efficient load transfer devices than joints which rely solely on aggregate interlock. A dowel bar jointed pavement will therefore offer a reduction in pavement thickness when compared with a plain sawn jointed pavement. The AASHTO pavement design method [8] allows a calculation comparing different joint designs to be undertaken. The calculation suggests that changing a dowel bar jointed pavement to a plain sawn jointed pavement will increase the required pavement thickness by approximately 25 mm. Similar calculations can be found in other pavement design manuals.

The American Concrete Pavement Association (ACPA) recommends [9] that non dowel bar jointed pavements should be restricted to a maximum traffic flow of 120 trucks per lane per day or a maximum pavement life of 5 msa. UK and European experience confirms that undowelled pavement designs are only suited to traffic levels up to 5 msa. The ACPA advice is based on a Minnesota Department of Transportation study [10], which investigated joint efficiency in relation to pavement performance and concluded that plain aggregate interlock joints can safely be used for traffic levels up to 80–120 trucks per lane per day. It is noted that a recently completed major road in Australia was constructed without dowel bars.

The ACPA paper [9] also records research confirming that high subgrade strength, large size and angularity of coarse aggregate within the concrete, high aggregate quality and thick slabs can all improve aggregate interlock joint efficiency. Close joint spacing is also noted as more efficient than larger spacing since this reduces variation in joint width due to concrete expansion and contraction.

9.4   Joint sealing

Pavements trafficked by highway vehicles are generally constructed with sealed joints. The seal is intended to:

- prevent the ingress of moisture;
- keep debris out of the joint gap;
- allow unrestricted movement in the joint.

If the joint becomes blocked and unable to close, the slab will not be able to expand in hot weather and can fail in the form of a compression or blow-up failure. Only in certain specific cases may pavement joints be left unsealed. For example, UK military airfields can be constructed with unsealed joints since the pavements are regularly swept and are very lightly trafficked. Unsealed joints are not recommended for sites subjected to highway trafficking.

A number of features are essential to the efficient operation of a sealed joint; Figure 9.3 illustrates each important issue. The joint must:

- include a de-bonding agent at the base of the seal;
- have edges carefully formed to create a chamfer or rounded profile;
- not be formed of overworked concrete; overworking will create a weakened concrete matrix, leading to spalling;
- be of a width designed such that the seal can accommodate the expected joint movement.

Joint sealants need to be designed to cope with the anticipated joint movement; URC joints will move relatively little but JRC joints, which are generally at much greater spacing, are subjected to significantly larger movements. Different seal widths and materials can be used in the different

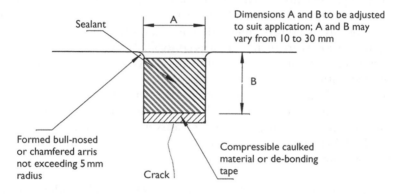

Figure 9.3 Joint sealant details.

Table 9.1 Joint sealant dimensions, UK Highways Agency [11]

	Contraction joints		Longitudinal joints	
	Width A (mm)	Depth B (mm)	Width A (mm)	Depth B (mm)
URC system	10	13	10	13
JRC, short slab system	13	15	10	13
JRC, long slab system	30	25	10	13
	Expansion and isolation joints			
	Width A (mm)		Depth B (mm)	
All systems	30		25	

types of joint. Table 9.1 summarises the UK Highways Agency sealant widths.

The joint dimensions given in Table 9.1 are based on hot poured material, joint widths may be reduced by 2 mm if cold poured systems are used.

9.5   Joint types

The following sub-sections describe the four different types of joint which may be formed in surface slab systems, each joint type performing a different function in the pavement system. Different joint designs and spacings are used in each of the major surface slab systems. It is noted that one of the best references describing the engineering functions of each joint type may be found in the ACPA papers [9,12] on joints.

Contraction joints

In a highway, these are transverse joints, linking together concrete panels along the line of trafficking. Standard practice is to build concrete pavements in continuous strips along the line of the carriageway for each individual lane. The transverse joints are then created as controlled cracks dividing each strip into rectangular or square panels. Joints may be constructed as dowelled or undowelled; different joint designs and spacings are used in URC and JRC pavements. On large expanses of concrete pavement, such as at airfields, contraction joints may also be specified longitudinally. The principles applying to such joints are exactly the same as for transverse joints although construction methodology is inevitably rather different. The essential features of a contraction joint are:

- When dowel bars are used they must be straight and bonded into one side of the joint.
- Dowels must be placed parallel to the line of the road and fixed at mid-depth in the slab.
- In the UK, dowels are usually 500 mm long in grade 250 plane bar at 600 mm spacing; German practice is to use a tighter 300 mm spacing under wheel tracks.

Figure 9.4 illustrates a number of different standard joint types.

Longitudinal warping joints

Longitudinal warping joints link together continuous strips of formed concrete slab. The joint performs a different function from that of transverse joints and is designed to different engineering principles; essentially these joints tie together strips of URC, JRC or CRC pavement, allowing a degree

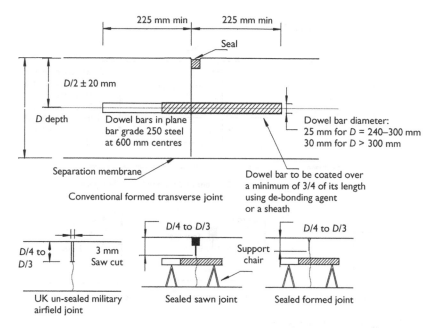

Figure 9.4 Transverse joint types.

of bending (as individual slabs warp due to temperature effects) but not opening or closing. The joints are constructed using steel tie bars. The essential features of a longitudinal warping joint (see Figure 9.5) are:

- Tie bars are fixed at mid-depth of the slab and bonded to either side of the joint. The central section of the bar is given corrosion protection.
- Tie bar specification is a function of slab type and thickness; 750 mm long, 12 mm diameter deformed bars of Grade 460 steel at 600 mm centres form a frequently used standard design although other permutations are also encountered.
- Joints may be formed or crack-induced. Crack inducers are used in the same manner as for transverse contraction joints.
- No more than three parallel longitudinal warping joints are recommended; on wide pavements such as airfield aprons, every fourth longitudinal joint should be a contraction joint, as described in the Section 'Contraction joints', in order to allow for expansion and contraction effects.
- European practice is to carefully arrange longitudinal joints to avoid wheel path locations.

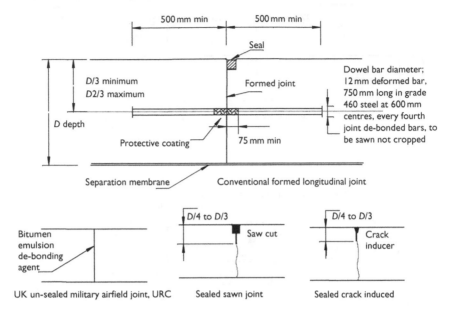

Figure 9.5 Longitudinal joint types.

Expansion joints

Expansion joints are generally transverse joints constructed with an allowance for expansion of adjacent slabs as a means of preventing compression failures or blow-ups in JRC and some URC pavements. The appropriate use of expansion joints is important; they should always be used sparingly since they lead to a reduction in joint efficiency. An ACPA paper [9] describes how expansion joints should best be used. Expansion joints tend to close up with time leading to an opening out of other transverse joints. The opened out joints may then contribute to sealant failure, water infiltration and loss of aggregate interlock. Unnecessary specification of expansion joints should therefore be avoided. They are usually only needed in long slab JRC systems or with materials that are known to expand. The UK Highways Agency recommends that expansion joints should only be used when:

- the slab is constructed when ambient temperature is below 4°C.
- the pavement is constructed of materials that have shown high expansion characteristics.

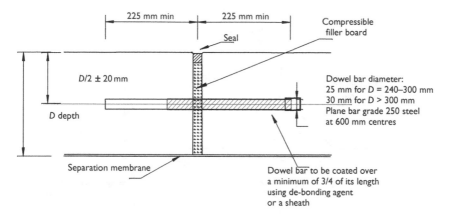

Figure 9.6 Expansion joint.

The essential features of an expansion joint (see Figure 9.6) are:

- Compressible filler is provided between slabs.
- The joint must be dowelled with caps on one end of the dowel to allow expansion.
- The dowel must be half embedded and fixed in a slab centre to the same specification as transverse joints.
- A seal is needed; the seal is frequently constructed to a minimum width of 30 mm and depth of 20–25 mm.

Isolation joints

Isolation joints are required at the interface between rigid stiff objects and the main pavement slab. They are intended to allow differential horizontal or vertical movement between the concrete slab and such objects as manholes, structural columns, concrete drains, etc. The joints are used to isolate these objects from the main pavement structure. The main features of an isolation joint are similar to those of an expansion joint except that they generally exclude dowel bars. Figure 9.7 illustrates a typical example around a manhole structure.

9.6   URC joint detailing

In URC pavements, joint spacing is critical to the efficient operation of the pavement, although different recommendations may be found in different design standards (see Table 9.2). US airfields advice suggests that the most

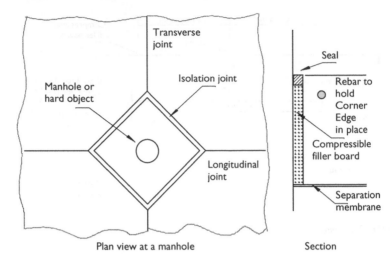

Figure 9.7 Isolation joint.

Table 9.2 Summary of joint spacing requirements

Concrete thickness (mm)	Transverse joint spacing			Movement joint spacing (m)
	ACPA [12] (m)	UK highways[a] [13] (m)	UK airfields[a] [4] (m)	
150	3.7–4.6	4	3	18
200	4.6	4	3	22.5
250	4.6	5	5.26	None
300	—	5	6	None

Note
a All UK standards allow joint spacing to be increased by 20% if limestone aggregate is used.

successful pavement performance is achieved with joint spacing between four and six times the radius of relative stiffness (see Chapter 5) of the pavement system. Various investigations are recorded in an ACPA paper [9] which recommends that maximum efficiency is achieved by a joint spacing of 25–30 times the pavement thickness, up to a maximum spacing of 4.5 m. UK Highways Agency advice allows slightly wider spacing; if limestone aggregate is used, the spacing may be increased by 20% since limestone has a much lower coefficient of thermal expansion than other minerals. Longitudinal joint spacing depends largely on the transverse spacing. ACPA [9] suggests spacing that is a factor of 1.25–1.5 less than the transverse spacing. Generally, it is advised that slabs should be square at the maximum permitted spacing

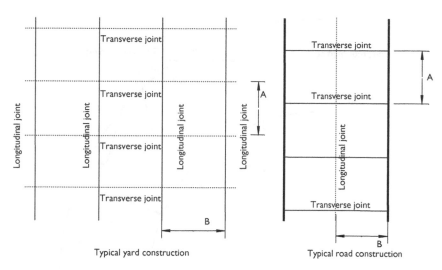

Figure 9.8 URC joint layout.

but, if a rectangular panel is used, a maximum aspect ratio of 1.5 is prudent in order to avoid warping in one direction becoming dominant, leading to concentration of stress; in exceptional cases the ratio may be stretched to 2 before reinforcement is added. Figure 9.8 illustrates a typical jointing pattern.

A specific problem exists with construction around a circular bend, common in airfield construction. For this, it is recommended that, if possible, transverse joints are maintained in a radial pattern around any bend system. Figure 9.9 illustrates a typical circular curve layout.

9.7 JRC joint detailing

Two different approaches may be undertaken to designing JRC slab systems; either a large or a small slab system may be used. US practice is described in the ACPA paper [9] and utilises a short slab system. The paper records the following details:

- maximum joint spacing 9.5 m;
- maximum longitudinal joint spacing taken from the URC pavement method;
- no expansion joints are to be used in the pavement.

UK practice [13] utilises a long slab system where joints are fixed at different rates for different reinforcing standards. The standard also indicates

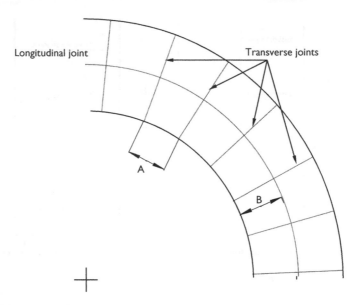

Figure 9.9 A circular bend jointing layout.

Table 9.3 JRC joint spacing

Slab thickness (mm)	Transverse (m)	Longitudinal (m)	Expansion (m)
Short slab system			
200	10	6	None
250	10	6	None
Long slab system; 500 mm^2 reinforcement or more			
200	30	6	90
250	25	6	75
300	23	6	69

Note
Transverse joint spacing may be increased by 20% if limestone aggregate is used.

that joint spacing may be increased by 20% if limestone aggregate is used. The standard suggests that maximum spacing is a function of slab thickness and level of reinforcement. The latest version of the standard gives transverse joint spacing for a pavement with 500 mm^2 of reinforcement per metre length but does not offer any advice on expansion joint frequency. UK historic practice is to fix expansion joints at every third transverse joint. Table 9.3 contains the best advice the authors can give, based on their experience.

Table 9.4 CRC system joint spacing

Slab thickness	Longitudinal joint spacing	
	All aggregates	Limestone
All slab thicknesses	6 m	7.2 m

9.8 CRC joint detailing

CRC slab systems require careful construction of longitudinal joints. If the system is constructed as a continuous slab without longitudinal joints the slab will warp and crack naturally at the point where the joint should have been formed. The UK standard recommends forming longitudinal joints (described in Table 9.4), at a maximum of 6 m centres; increasing transverse reinforcement will not remove the need for longitudinal joints.

9.9 Strong CBM jointing

Two alternative strategies may be used in the construction of CBM base or sub-base layers; the material can be allowed to crack naturally or cracks can be induced into the layer. Each technique is described separately.

Natural cracking in CBM materials

US roller compacted concrete practice has been to allow naturally formed cracks to occur in a machine-laid layer. Historically a similar practice has been followed in the UK using 'lean-mix' and strong cemented materials. The material was allowed to crack naturally into large blocks; the shrinkage cracks then became transverse cracks and the machine-laid strip edges longitudinal joints. The Portland Cement Association [2] observes that a typical naturally occurring crack spacing will seldom exceed 20 m and some schemes have been noted as cracking at as little as 4 m centres. The closer cracking tended to be associated with schemes over 10 years old. Naturally formed cracks are rarely sealed deliberately; debris from the pavement surface can frequently be found 'sealing' the joint. The wider naturally occurring shrinkage cracks are noted in [2] as producing reasonably efficient joints. This report observes that little evidence of faulting can be found. Joints are reported as having a typical efficiency of 60–90%.

Pre-cracking of CBM Materials

The precise nature of shrinkage-induced cracking is a function of: cement type and content, water–cement ratio and the environmental conditions present as the concrete matures. Current UK practice is to induce cracks within a CBM either as the material is laid or by 'guillotining' the recently set material. The induced cracks then form a regular transverse joint pattern. Normal practice is to space cracks at 3 m centres in any material with a compressive cube strength over 10 MPa at 7 days. Induced cracks can be formed as slots by using a fin attached to the base of a vibrating plate, with a depth of between 1/4 and 1/3 of the slab thickness. These slots are then filled with a bituminous emulsion prior to the layer being rolled. The rolling action closes the slots but the presence of bitumen ensures that a weakness remains, forming a preferential location for crack initiation. Guillotined joints are formed with a large purpose-made falling edge plate machine, the impulse from which fractures the slab, inducing hairline cracks into the CBM material.

9.10   Conclusions

Joints are absolutely essential to the efficient operation of a concrete pavement. However, the precise detailing of any system must be focused around the intended pavement design method and expected function.

9.11   References

1. The Highways Agency, *Manual of Contract Documents for Highway Works, Volume 7 Pavement Design and Maintenance*, section 2 Pavement Design and Construction, HD 29/94.
2. Piggott, P.E., Roller compacted concrete pavements – a study of long term performance, Portland Cement Association, March 1999.
3. Tayabji, S., Performance of continuously reinforced concrete pavements in the LTPP program, 7th International Conference on Concrete Pavements, September 2001, Florida.
4. Property Services Agency, *A Guide to Airfield Pavement Design and Evaluation*, HMSO, 1989, ISBN 0 86177 127 3.
5. Hanekom, A., Horak, E. and Visser, A., Aggregate interlock load transfer efficiency at joints in concrete pavements during dynamic loading, 7th International Conference on Concrete Pavements, September 2001, Orlando, Florida.
6. Weaver, J., The effect of dowel bar misalignment in the joints of concrete roads, Cement and Concrete Association, November 1970.
7. Weaver, J., A comparison of some bond preventing compounds for dowel bars in concrete road and airfield joints, Technical Report 403, Cement and Concrete Association, June 1967.
8. American Association of State Highway and Transportation Officials, *AASHTO Guide for Design of Pavement Structures*, American Association of State Highway and Transportation Officials, 1992, ISBN 1-56051-055-2.
9. American Concrete Pavement Association, *Proper use of isolation and expansion joints in concrete pavements*, ACPA, 1992, IS400.01P.

10. Review of Minnesota's concrete pavement design, Concrete Design Task Force, Minnesota Department of Transport, March 1985.
11. The Highways Agency, *Manual of Contract Documents for Highway Works, Volume 1 Specification for Highway Works*, section 1000, May 2001, London.
12. American Concrete Pavement Association, *Design and Construction of Joints for Concrete Streets*, ACPA, 1995, IS061.01P.
13. The Highways Agency, *Manual of Contract Documents for Highway Works, Volume 7 Pavement Design and Maintenance*, section 2, Pavement Design and Construction, TD 26/01, 2001, London.

Chapter 10

Detailing

10.1 Introduction

Concrete and cement bound layers obviously require quite different detailing from crushed rock and bituminous materials. Equally obviously, the successful detailing of a pavement is essential to a maintenance-free life. Many instances of poor ride quality and serviceability are due to badly built or specified pavements. A high-quality concrete pavement requires appropriate detailing and this chapter discusses the features that are essential to production of such a pavement.

10.2 Minimum layer thickness

The first most important issue to consider associated with the detailing of concrete pavements is the minimum layer thickness. Construction tolerances, site plant and the practical implications associated with building pavements dictate a minimum and in some cases maximum limit on the thickness of layers in the pavement. The following limits to thickness are suggested based on UK construction experience.

CBM materials

Minimum recommended layer thickness   150 mm
Maximum recommended layer thickness   200 mm

Thick CBM bases are achieved by building up multiple layers of material. CBM materials are laid semi-dry. The maximum layer thickness is suggested as the limit that conventional construction plant can successfully complete. If the layers are thickened beyond 200 mm specialist compaction advice will be required or the CBM material may need to be laid in multiple layers. Ideally, the second layer in a two-layer system is placed immediately over the first layer, before the first layer has set. The minimum 150 mm layer thickness is suggested as the limit for which construction tolerances will allow a reasonable chance of success. Any thinner and errors in the subgrade level, which can frequently be in the order of ±30 mm, will produce an unacceptable reduction in the total CBM layer thickness. An excessively thin CBM layer is susceptible to site traffic damage, which greatly reduces the effectiveness of this type of design.

URC slabs

Minimum recommended thickness   150 mm
Maximum recommended thickness   500 mm

A similar minimum layer thickness of 150 mm is recommended for mass concrete slabs. The minimum thickness is again suggested as reductions in the slab depth, resulting from errors due to foundation construction tolerance, adversely affect the ultimate pavement strength. A planned 150 mm slab with a construction tolerance of ±20 mm at top of foundation and ±10 mm at the finished pavement surface could, in some exceptional locations, produce a reduction of 30 mm in the slab thickness. This excessively thin slab will clearly be substantially weaker than the planned, calculated thickness. The maximum slab thickness of 500 mm is suggested based on UK airfields experience, where slip-formed slabs of this thickness have been constructed at heavily trafficked military and civil airfields.

JRC and CRC slabs

Minimum recommended thickness   200 mm
Maximum recommended thickness, single layer rebar   300 mm
Maximum recommended thickness, double layer rebar   unknown

Singly reinforced concrete slabs, with the reinforcement laid at mid-depth in the slab, are currently constructed in the UK to a minimum thickness of 200 mm. Building slabs to a layer thickness thinner than 200 mm produces problems with ensuring an efficient and consistent cover to the reinforcement. Reinforcement fixed in slabs thinner than 200 mm can often deviate from mid-depth producing inconsistent construction. The maximum single layer reinforced slab thickness is based on US highway experience where thick slabs are reported as failing in shear in the plane of the reinforcement. It is suggested that slabs thicker than 300 mm might be more effectively constructed with reinforcement both top and bottom.

10.3 Surface finish

Background

The surface finish of a pavement system requires special attention and detailing to ensure an effective pavement surface is produced. Many different techniques are used in association with various construction methods. This section gives advice on the most appropriate techniques for each type of construction. The essential features are:

1 The surface must be stable.
2 Noise generation characteristics must be minimised; a maximum surface texture of 1.25 mm is recommended in UK standards.
3 The ride quality must be acceptable.
4 The skidding characteristics must be considered.

A comprehensive design note may be found in a PIARC report [1] which recommends that surface characteristics may be properly designed if each of the following surface texture effects, occurring at different wavelengths, are considered:

The PIARC committee reported that each design issue can be addressed by provision of essential pavement characteristics as follows in Tables 10.1 and 10.2.

The PIARC report is a good general source of data but a second reference is recommended describing the relationship between concrete aggregate and skidding resistance. A state of the art review, Road aggregates and skidding [2] by Hosking indicates that fine aggregate substantially influences a concrete pavement's skidding characteristics. Coarse aggregate was reported as having a more minor influence on skid resistance. The other

Table 10.1 Summary of surface characteristics

Type of surface property	Range of dimensions (mm)		Pavement surface characteristic	Influences
	Horizontal	Vertical		
Microtexture	0–0.5	0–0.2	Surface texture of aggregate or groove Edges	Skid resistance High frequency tyre noise
Macrotexture	0.5–50	0.2–10	Aggregate size Surface treatment characteristic	Skid resistance (wet) Spray High and low frequency tyre noise Rolling resistance
Megatexture	5–50	1–50	Regularity of pavement surface	Vehicle control Rolling resistance Comfort Tyre and vehicle wear Low frequency rumble
Roughness	500 mm–500 m	1–20	Level control	Comfort Vehicle control Fuel economy Vehicle wear

Table 10.2 The influence of surface finish on ride quality [1]

Characteristic	Texture			Roughness
	Micro	Macro	Mega	
Skid resistance	a	a		
Road-holding qualities				b
Splash and spray		a		
Reflectance		a		
Dynamic loads				b
Vehicle wear				b
Tyre wear	b			
Rolling resistance		b	b	(b)
Vibrations (inside vehicles)			(b)	b
Noise (inside vehicles)			b	
Noise (outside vehicles)		a	b	

Notes
a Essential feature for effective pavement surface.
b Undesirable feature.

issues influencing skid resistance were reported as:

- sand content;
- concrete strength;
- hardness of the coarse aggregate; softer aggregate gave better results.

This report [2] is recommended for any engineer considering the design of a major high-speed concrete pavement.

Low-speed finishes

Trowel

The pavement surface is finished with a trowel to produce an even, sealed surface, this type of finish is only used on untrafficked or internal slabs. The surface will be without sufficient texture to give a reasonable wet skidding resistance.

Rough tamp

The pavement surface is completed with a hand controlled tamp giving a system of random ripples of between 0 and 3 mm in height. The system is only suitable for lightly trafficked sites, with low-speed vehicles.

Hessian drag

Another low-speed road finishing technique [3] involves the use of a sheet of hessian, typically 0.6 or 1.2 m long, which is dragged longitudinally over the plastic concrete. The 0.6 m system gives a 'light hessian drag', the 1.2 m system a 'coarse hessian drag'. The finish is adequate for low-speed roads.

Broom finish

A useful finish can be produced by dragging a light broom over the finished pavement surface this method is appropriate for low speed, factory yard or estate roads. The broom is dragged across the pavement surface at 90° to the direction of traffic producing an even regular texture to the finished concrete.

High-speed highway finishes

Chipping the surface

This system involves the application of high polished stone value (PSV) chips of size 10–14 mm or 14–20 mm at the rate of 6–7 kg/m^2. The method is described in references [3] and [4] and has been used in France and Belgium. The chips are mechanically spread and vibrated or rolled into place. The references suggest that the system has met with mixed success; chips are reported to pluck and become detached from the pavement, while the positive texture produces a noisy surface.

Coarse aggregate exposure

A thin 50 mm layer of specially batched concrete, with 7–8 mm coarse aggregate, is placed over the main pavement slab (while it is still wet). The thin layer is therefore monolithic with the main pavement. The surface is sprayed with a retarding agent and, following initial set, the surface fine aggregate is removed with a brush to expose the negative texture of the coarse aggregate. The technique requires extensive skill and is consequentially difficult and expensive. The method is also described in references [3] and [4].

Tined finish

A randomised system of indents may be used and is reported in [5] as a technique historically used as a pavement finish. The paper reports using 2 or 3 mm deep indentations with the following repeated spacing pattern (in mm):

10, 14, 16, 11, 10, 13, 15, 16, 11, 10, 21, 13, 10.

The Australian paper [5] reports that a light hessian drag combined with a 2 mm tined finish gives the optimum surface finish. A similar technique is recommended in the UK highways specification; the standard uses a groove depth between 2 and 5 mm with the following randomised pattern (spacings in mm):

40, 45, 35, 45, 35, 50, 30, 55, 35, 30, 50, 30, 45, 50, 30, 55, 50, 40, 35, 45, 50, 40, 55, 30, 40, 55, 35, 55.

The random pattern is needed to avoid noise generation problems.

Open textured bituminous wearing course

A successful technique is used in the UK and Holland whereby a 10 or 14 mm open, negative textured, wearing course is placed over the main structural slab. The Dutch technique uses a 50 mm porous Asphalt held in place using a bituminous bond coat; the UK technique uses a 35 mm layer of patented 14 mm open textured wearing course. Both systems work well and produce pavements that are easily maintained. The porous Asphalt produces a high quality, low noise, surface similar to the exposed aggregate surfaces but without the difficulties experienced in laying a specially batched concrete layer over the pavement. The system seems to be currently confined to Europe but is likely to be used extensively throughout the world in the near future.

Recommendations

The recommended pavement surfaces for different applications are:

1   Low-speed roads, industrial sites and other low risk situations

   - Brush or hessian drag.

2   High-speed major highways.

   - CRC   open textured bituminous wearing course overlay;
   - URC, JRC   tined finish – although relatively high noise still expected.

Figure 10.1 summarises the different surface finishing techniques.

10.4   Slip membranes

Background

Plastic sheet slip membranes, 125 microns thick, are frequently used under URC and JRC slab systems. The plastic sheeting is intended to allow the

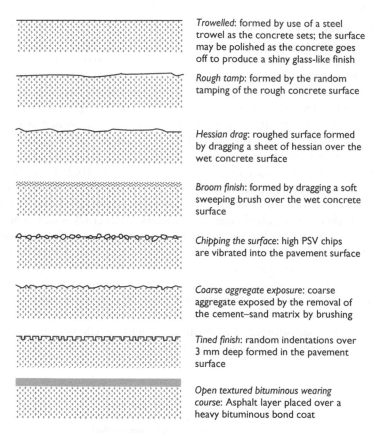

Trowelled: formed by use of a steel trowel as the concrete sets; the surface may be polished as the concrete goes off to produce a shiny glass-like finish

Rough tamp: formed by the random tamping of the rough concrete surface

Hessian drag: roughed surface formed by dragging a sheet of hessian over the wet concrete surface

Broom finish: formed by dragging a soft sweeping brush over the wet concrete surface

Chipping the surface: high PSV chips are vibrated into the pavement surface

Coarse aggregate exposure: coarse aggregate exposed by the removal of the cement–sand matrix by brushing

Tined finish: random indentations over 3 mm deep formed in the pavement surface

Open textured bituminous wearing course: Asphalt layer placed over a heavy bituminous bond coat

Figure 10.1 Concrete pavement surface finishes.

concrete slab to move freely over the support platform, which is necessary in order to accommodate thermal expansion and contraction effects. However, an alternative technique is frequently used in the UK when pavements are constructed over CBM bases, or for CRC pavements. A bituminous-sprayed membrane, using K1-40 emulsified bitumen at the rate 0.6 l/m^2, is sprayed over the CBM layer as a curing membrane. The plastic sheet produces an unbonded slab system; the bituminous spray produces a partially bonded system. Chapter 5 (the Sections 'Adjacent slab support' and 'Environmental effects') demonstrates that a two layer bonded system will be stiffer than a two layer unbonded system. The calculation would suggest that placing a plastic sheet membrane between a CBM support platform and a strong surface slab system produces a weakened pavement.

Slip membranes and crack-induced joints

The authors consider that best practice on sites where a crack-induced URC or JRC pavement is used is to apply the K1-40 tack coated system in preference to a plastic membrane. The bituminous tack coat provides a partial bond between the cement bound construction platform and the concrete pavement slab. When the regular transverse joints are formed, by crack inducing, the partially bonded system offers a restraining force to the concrete slab. This restraining force ensures that each individual transverse contraction joint cracks and opens approximately the same amount. If a plastic sheeting membrane is used the slabs tend to move and crack in an unrestrained, irregular manner, as illustrated in Figure 10.2. Typically each third joint will crack and open, leaving the intermediate joints to crack under the action of traffic at a later date. A pavement constructed over a plastic membrane therefore results in uneven transverse joint crack width. The plastic sheet system can lead to joint widths of between 0 and 5 mm, whereas it has already been noted that a joint width of less than 1 mm is needed to ensure efficient aggregate interlock across a joint. Wide joints result in a loss of load transfer efficiency and a concrete pavement with poor joint load transfer characteristics is substantially weaker than a well-constructed pavement.

Against this background, the existing recognised UK specifications for concrete pavements are considered suspect.

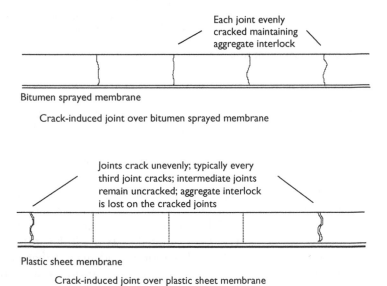

Figure 10.2 The impact of slip membranes on crack-induced joints.

Recognised UK specifications

Two distinct standard specifications are currently used for rigid concrete slabs in the UK; the recognised texts are:

- Highway work: Highways Agency's *Specification for Highway Works* [6]. Clause 1007: *Separation and waterproofing membranes.* The clause has recently undergone a revision, and it now insists that an impermeable plastic sheeting membrane, 125 microns thick, must be used under all URC and JRC pavements. A bituminous-sprayed tack coat system is required under CRC systems.
- Airfield work: Defence Works Function Standard, Specification 033, *Pavement Quality Concrete for Airfields* [7]. Slip membranes are detailed under Clause 3.8: *Separation Membrane.* The clause is mandatory, insisting on the use of a 125-micron thick impermeable polythene sheeting.

The Highway Agency's specification [6] was revised in May 2001, changing the clause from a position where either the bituminous tack coat or the plastic sheeting system could be used under concrete pavements.

Recommendations

The authors' current best advice is to use bituminous-sprayed membranes in place of plastic sheeting for the reasons given in the Section 'Slip membranes and crack-induced joints'. The bituminous-sprayed membrane can be constructed to Clause 920 of reference [6].

10.5 Highway carriageway rollovers

The construction of rollovers, the transfer of cross-fall from one side to the other, is a difficult detail on large multi-lane carriageways and rollovers constructed in concrete slab systems are noted as requiring special attention. The most efficient method of construction is to roll each individual lane in turn; this method tends to avoid the creation of flat spots. However, the complex level control needed to build this layout inevitably leads to the hand laying of this section of the pavement.

10.6 Transitions between different types of construction

An important design consideration for concrete slab systems is to produce a correctly designed transition or rolling block between different types of construction. It is also noted that a ripple will form in an adjoining bituminous pavement if a movement joint (i.e. an expansion joint) is not provided

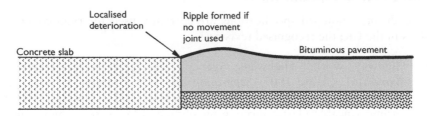

Figure 10.3 Effect of omitting movement joint between flexible and rigid construction.

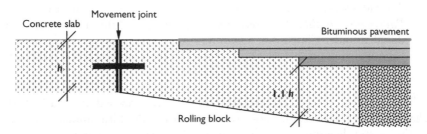

Figure 10.4 Transition from rigid to flexible construction.

near the change from flexible to rigid construction. Figure 10.3 describes the problem. The ripple is formed as a consequence of the different thermal behaviour induced in each pavement type. The rolling block minimum pavement thickness should be 1.1 times the main pavement thickness to avoid any chance of cracking. Figure 10.4 describes a typical arrangement.

10.7   Slip road connections into a main carriageway

A specific detailing problem exists in connecting slip roads into a concrete surface slab system. UK practice is to construct the slip road widening and nosing in the same construction as the main carriageway. The slip road is then formed as a widened-out section of the main carriageway with a transition slab at the end of the nosing. Figure 10.5, from the UK standard, illustrates typical arrangements. It is noted that constructing the slip roads in this way usually requires hand laying of the complex widened-out area.

10.8   CRCP anchorages

Anchorages are essential to the efficient operation of a CRCP pavement. However, the design and detailing of anchorages is little understood and the result is that most pavement designers use a system of custom and practice. If a pavement is constructed without anchorages or movement joints the

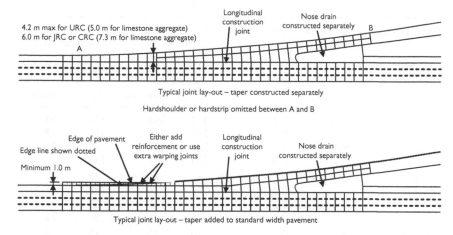

Typical joint lay-out – taper constructed separately

Hardshoulder or hardstrip omitted between A and B

Typical joint lay-out – taper added to standard width pavement

Figure 10.5 Connection of slip roads into a main carriageway.

pavement ends will move very substantial distances in response to thermal effects. Very little engineering information is published on the design of anchorages and thermal movement but one of the best summaries of the problems may be found in a paper by McCullough and Moody [8].

Unrestrained movement

A number of different factors contribute towards the magnitude of thermal movement in a pavement:

- aggregate type;
- pavement thickness;
- environmental temperatures when the pavement was constructed;
- support platform type, material and slip membrane.

McCullough and Moody note that a pavement will initially resist a thermally induced stress in the pavement until the coefficient of friction with the foundation is overcome. Once the pavement begins to move, the frictional drag is reduced and the pavement will move relatively freely until it stops and the drag must again be overcome. Table 10.3 reports the seasonal movements from the paper.

The data is based around the following conditions:

- summer pavement construction temperature 31°C;
- winter pavement construction temperature 21°C;
- winter air temperature 2°C to 12°C;

Table 10.3 Seasonal movements in unrestrained CRCP [8]

Foundation type	Placement season	Slab thickness (m)	Max mobilised length (m)		Seasonal movement (mm)	
			Aggregate		Aggregate	
			Gravel	Limestone	Gravel	Limestone
Plastic sheeting	Summer	0.2	304	243	145	94
on crushed		0.3	380	334	218	138
rock	Winter	0.2	228	182	98	61
		0.3	304	243	145	90
Bituminous	Summer	0.2	61	61	33	21
Asphalt		0.3	91	76	48	31
	Winter	0.2	61	61	23	16
		0.3	76	76	33	22
Cement	Summer	0.2	30	30	12	8
stabilised base		0.3	46	30	17	10
	Winter	0.2	30	30	9	7
		0.3	30	30	12	9

- summer air temperature 29°C to 40°C;
- autumn air temperature 13°C to 26°C.

The study clearly demonstrates the importance each factor can have in influencing the magnitude of force or movement that needs to be carried into or restrained by an anchorage. The Texas research [8] also usefully includes a prediction model. The model has been converted to metric for the purposes of this book and is given in Equation 10.1.

Equation 10.1 To predict unrestrained movement in CRCP

$$\text{End movement} = \text{Ses}\frac{h}{0.2}\Delta T \left(\frac{\alpha}{6 \times 10^{-6}}\right)^{1.5} f \tag{10.1}$$

Ses is the season of placement; 1.33 for summer, 1.0 winter
h = concrete slab thickness (m)
α = 'thermal coefficient', expressed in strain per °C
f = sub-base frictional factor; 0.037 plastic sheeting, 0.0086 Asphalt base, 0.0031 cement treated base

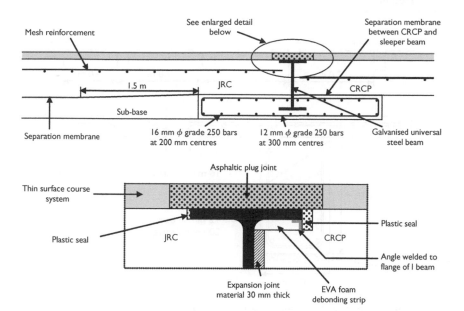

Figure 10.6 UK standard I-beam termination system.

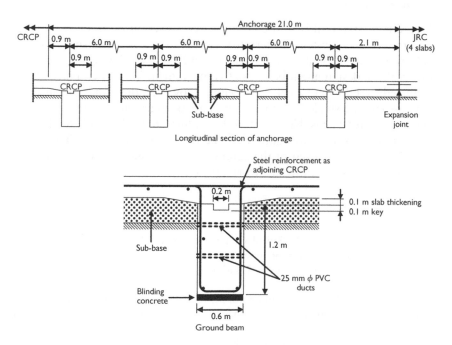

Figure 10.7 UK reinforced concrete ground beam details.

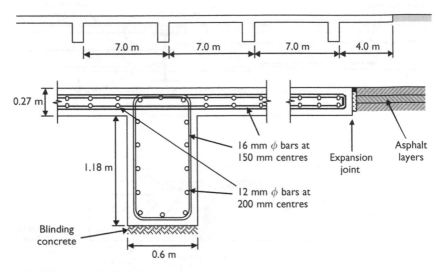

Figure 10.8 Dutch reinforced concrete ground beam system [9].

Design solutions

The industry standard is to use either a steel I-section beam or an in situ-cast ground beam system for terminations. The Texas paper suggests that anchorages may be omitted if movement can be reduced below 50 mm but a more realistic design might be to use two movement joints if end movement can be reduced down to a value of 10 mm. However, the standard solutions are to use either a single I-section beam or an arrangement of four reinforced concrete ground beams. Figures 10.6–10.8 illustrate the UK and Dutch standard details. No recommendation can be made commending either design over the other, since each of the solutions can be successfully constructed and both systems are known to work efficiently.

10.9   References

1. Surface characteristics, Technical Committee Report No. 1 P.I.A.R.C., World Road Congress, Brussels, September 1987.
2. Hosking, R., Road aggregates and skidding, *State of the Art Review 4*, HMSO, London.
3. Charonnat, Y., Lefebvre, J. and Sainton, A., Optimisation of non-skid characteristics in construction of concrete pavements, 4th International Conference on Concrete Pavement Design and Rehabilitation, April 1989, Purdue University.
4. Descornet, G., Fuchs, F. and Buys, R., Noise-reducing concrete pavements, 5th International Conference on Concrete Pavement Design and Rehabilitation, April 1994.

5. Nichols, J. and Dash, D., Australian developments to reduce road traffic noise on concrete pavements, 5th International Conference on Concrete Pavement Design and Rehabilitation, April 1994.
6. Highways Agency, *Manual of Contract Documents for Highway Works, Volume 1 Specification for Highway Works.*
7. Defence Estate Organisation, *Specification 033, Pavement Quality Concrete for Airfields*, ISBN 0 11 772476 9.
8. McCullough, B.F. and Moody, E.D., *Prediction of CRCP Terminal Moments*, 5th International Conference on Concrete Pavement Design and Rehabilitation, April 20–22, 1993, Purdue University.
9. Stet, M., *CRCP: A Long Lasting Pavement Solution for Todays Motorways*, Dutch Practice, 7th International Conference on Concrete Pavement Design and Rehabilitation, September 2001.

Index

Printed and bound by CPI Group (UK) Ltd, Croydon, CR0 4YY

01/11/2024

01782621-0004